Neuroscientific Basis and Purpose of Self, Mind, Consciousness and Emotion

Tomoo Hirano
Department of Biophysics
Graduate School of Science
Kyoto University, Japan

CRC Press
Taylor & Francis Group
Boca Raton London New York

CRC Press is an imprint of the
Taylor & Francis Group, an **informa** business
A SCIENCE PUBLISHERS BOOK

Cover credit: Images reproduced by kind courtesy of the author, Tomoo Hirano.

First edition published 2026
by CRC Press
2385 NW Executive Center Drive, Suite 320, Boca Raton FL 33431

and by CRC Press
4 Park Square, Milton Park, Abingdon, Oxon, OX14 4RN

CRC Press is an imprint of Taylor & Francis Group, LLC

Library of Congress Cataloging-in-Publication Data (applied for)

ISBN: 978-1-032-83649-2 (hbk)
ISBN: 978-1-032-85579-0 (pbk)
ISBN: 978-1-003-51882-2 (ebk)

DOI: 10.1201/9781003518822

Typeset in Times New Roman
by Prime Publishing Services

Preface

In this book I aim to provide a systematic and unified view of self, mind and consciousness, or a grand framework of the mind and the brain, based on neuroscientific and biological findings. The book discusses how and why the self, mind, consciousness and emotions have emerged. The basic idea is as mentioned in the following paragraphs.

Animal species have needed to survive during evolution, and one strategy for survival is to enhance adaptability to the environment with the help of an advanced information processing system. Towards this end, some animal species have developed their nervous systems so that they can gather and process various sensory signals, learn and remember experiences; and then consider the obtained information for decision making, selection of behavior and making action plans. These highly intelligent processes are useful in the survival competition and are performed with the nervous system consisting of neurons and synapses, whose functions are enabled by biochemical materials, such as various types of protein. These biochemical and biophysical processes have been objectively and intensively studied in neuroscience. During the evolutionary development of the animal species, self, mind and consciousness seem to have emerged.

One very difficult question, which has been called a hard problem, is how the objective physicochemical processes performed by neurons and synapses are related to the subjective mind and consciousness. The latter are absolutely personal, and strictly speaking one can only make a guess regarding another individual's mind and consciousness by an introspective consideration of one's own mind and experiences. This long-standing question is hard to resolve, if not impossible, because subjective experience is owned only by oneself and cannot be directly shared with other people, although conscious experiences can be explained by words or expressed by behaviors. I feel that it may not be wise to directly tackle the hard problem, and that another approach to learn about relations between objective neuroscientific phenomena and subjective experiences which are expressed verbally may be more fruitful. By sorting out such relations, I think, we can obtain a better understanding of the nature and characteristics of subjective existence and experiences, and also clarify their dependence on the objective materials such as the nervous system and sensory organs.

I also compare the functions of the brain with a computer system or artificial intelligence, as I think that such a comparison will help us to get a better understanding of characteristics of the neuronal networks of the brain. Although both of them are engaged in information processing with learning ability, there are certain differences. Their purposes and characteristics are discussed, taking whether each of them is a life or not into account. Emotions, dreams and neuropsychiatric disorders are also explained and discussed, because their understanding is useful to clarify the nature of self, consciousness and mind. I have tried to encompass related topics as far as I can. Through the above considerations, I have tried to provide a systematic and united view or a grand framework of self, mind and consciousness, although the unity of self and mind in an individual will be questioned.

Now, let me explain why I decided to write this book. Please allow me to tell you a little about my personal history. When I was a high school student in 1970s, I was interested in the philosophical question "What is myself?". However, I felt that the philosophical and/or psychological discussions which I read in books at that time were either enigmatic, or too speculative or not based on firm scientific facts. After I learned neuroscience, computer science, etc. as an undergraduate student in the university, I thought that one critical characteristic of the nervous system as compared with a computer system of those days, was flexibility or learning ability. Thus, I decided to study the basal mechanisms of learning in animals, which is synaptic plasticity. After getting a Ph. D., I chose cerebellar culture as a simple model experimental system to perform research on synaptic plasticity, and studied cellular and molecular mechanisms of cerebellar long-term depression, a type of synaptic plasticity underlying motor learning. Some reasons for the choice of the research theme were that the available neuroscientific research methods at that time were not satisfactory for me to tackle higher brain functions such as cognition, etc. in live mammals or humans, and that I wished to obtain some fundamental, critical and rigorous results that could be firm bases of the following neuroscientific research.

While I was working on cellular and molecular mechanisms of synaptic plasticity in 1990s in Kyoto University, neuroscientific techniques to generate transgenic and/or knockout mice had developed and behavioral experiments using mutant mice had become possible. Seeing such a situation, I became curious to examine the motor learning ability of mutant mice in which a critical molecule, necessary for induction of the cerebellar long-term depression, was knocked out. To this end, my collaborators and I built an experimental setup to examine the model motor learning of mice in the laboratory. I enjoyed these studies, which provided some fruitful results.

During and after those days, further advances in neuroscientific research methods, such as *in vivo* neuronal activity recordings using fluorescent dyes and optogenetic techniques which enabled the stimulation or inhibition of specific neurons *in vivo*, functional magnetic resonance imaging of human brains, non-invasive neuronal stimulation on the human brain, etc., have taken place. In addition, we have come to know more about the neuroscientific bases of higher brain functions, molecular mechanisms of some neuropsychiatric disorders, etc., which have made discussions about consciousness and higher brain functions popular. At the same time, computer technology, including artificial intelligence (AI) has also made impressive progress, and the utilization of AI has spread out to various areas. Now, if one cannot directly see the partner, it has become difficult to discover whether a partner in a conversation is a human or AI. Further, at some occasion we might feel that AI has its own mind. The above conditions make me reconsider "What is myself?".

There were also other reasons which prompted me to write this book. They were the social and political situations of universities, nations and human beings as a whole. I served as the Dean in the Graduate School of Science, Kyoto University from 2017 to 2019. During this period, I needed to understand and consider various matters and affairs in and around the graduate school and the university, such as their own merits and demerits in education and research, their financial status, the demands from the society and/or the government, etc. To handle various matters and affairs adequately, I began to take more interest in what was happening around the university, in Japan and in the world. Although I was a specialist in neuroscience research, I needed to be a generalist as a Dean. This experience made me consider various problems in human societies, such as conflicts among people and nations, wealth gap, handling of global warming, etc., and I began to feel that a better understanding of the nature of ourselves might be useful in helping us tackle various social problems.

In any case, here, I try to clarify subjective concepts such as consciousness, which has multiple meanings, and to explain the purpose of the living species and how and why the nervous systems have evolved. Then, I explain objective molecular, cellular and system level neuroscientific findings, which underly the subjective self, mind and consciousness, and how they are related. The similarities and differences of the brain and a computer system are discussed to clarify the distinct characteristics of the living brain. Emotions which seem to have evolved to mainly merit an individual and their closely related companions are also explained, and neuroscientific bases, self-centered characteristics, significant influences on decision making and possible demerits of emotions in a large society are discussed. Dreams and neuropsychiatric disorders, that are based on distinct and unusual states of the functional brain are also explained. I think that these aspects also shed light on the nature of the self, mind and consciousness. Finally, I try to relate all the above-mentioned facts and notions together so that a systematic and unified framework on self, mind and consciousness emerges. I hope that my trial can present a somewhat new, different and/or unexpected views on yourself and humans. After reading this book, if you become more interested in the nature of yourself and human beings, it will give me great satisfaction.

Acknowledgement

The author thanks Prof. S. Kawaguchi of Kyoto University for his constructive comments on the manuscript after carefully and critically reading the manuscript.

Contents

Preface iii

Acknowledgment vii

1. Self, Mind and Consciousness **1**

What is Myself? 1
Definitions of Consciousness 2
Wakefulness 2
Awareness 2
Qualia 3
Self-Recognition and Metacognition 4
Theory of Mind 5
Hard Problem 6
Neuronal Correlates of Consciousness (NCC) 7
Unconscious Processes 8
Summary and Conclusion 9
References 10

2. The Purpose of Life and the Role of the Brain **11**

What is Life? 11
The Survival of an Individual and a Species 11
Various Ways of Keeping a Species 12
Self, Offsprings and Groups 12
Self-Interest and Altruism 13
Recognition of Environment and Selection of Action 14
Reflex, Instinctive and Voluntary Actions 15
Emergence of The Consciousness and The Mind 15
Summary and Conclusion 17
References 17

3. Neurons and Synapses **19**

Elementary Units of The Nervous System 19
Electrical Information Transmission Using Membrane Potential 20
Ionic Channels and Pumps 23
Synaptic Transmission 24
Inhibitory Synapses 26
Transmitters and Receptors 27

Slow Modulation of Synaptic Functions and Neuronal Activities 28
Intracellular Signaling Cascade 29
Synaptic Plasticity 29
The Control of Transcription and Long-Term Modification of Neuronal Circuits 35
Summary and Conclusion 35
References 36

4. The Functions of Neuronal Circuits: Sensory Processing and Motor Control **39**
Peripheral and Central Nervous Systems 39
The Development of The Brain and The Spinal Cord 39
Simple Neuronal Pathways for Reflex 41
Sensation and Perception 42
The Peripheral Visual System 43
The Central Visual System 45
The Auditory System 48
The Somatosensory System 49
The Olfactory System 50
The Taste System 51
The Lower Motor Control System 52
Inhibitory Neuronal Pathways 52
The Higher Motor Control System 53
Homeostatic Regulation of Body States 54
Stress Response 55
Summary and Conclusion 55
References 57

5. The Neuronal Bases of Higher Brain Functions **59**
Higher Brain Functions: Perception, Recognition, Evaluation, Decision Making and Planning 59
Neuronal Coding 61
Binding Problem in Recognition 61
Conceptualization and Words 63
Sequential and Parallel Information Processing 63
Working Memory 64
Rhythm 64
Loops 65
Attention 66
Weighing of Information 66
Relationship among Attention, Perception, Recognition, Binding and Working Memory 67
Conflict in Perception and/or Recognition 67
Top-down and Bottom-up Signals 69
Mirror Neurons 69
Decision Making and Planning 70

Summary and Conclusion 70
References 72

6. Learning and Memory **74**
Three Types of Learning 74
Self-Organization 74
Critical Period and Experience-Dependent Maturation of Neuronal Circuits 76
Supervised Learning 76
Cerebellum and Motor Learning 76
Reinforcement Learning 78
Bayesian Estimation 79
Efficient Learning vs Flexibility 79
Born and Raised 80
Declarative and Non-declarative Memory 81
Engram 82
Summary and Conclusion 85
References 86

7. A Comparison of The Brain and A Computer System **89**
Computer System and Artificial Intelligence 89
Similarities of the Brain and a Computer 91
Differences between the Brain and a Computer 92
Diverse and Complicated Neuronal Circuits 93
Efficient Information Processing with Slow Elements in The Brain 94
Contrast Enhancement 94
Filling-in 95
Illusion 95
Information Transmissions with Line Connections and with Spreading Factors 96
Challenge for a Change by the Brain 96
The Significance of Diversity for The Survival of The Species 97
Singularity 98
Summary and Conclusion 98
References 99

8. Emotions and Feelings **101**
What are Emotions and Feelings? 101
Fear and Anxiety 102
Anger and Aggression 104
Disgust and Avoidance 105
Sadness and Regret 106
Hate 107
Pain 107
Pleasure and Reward 107
Love and Attachment 109
Empathy, Sympathy and Compassion 109
Envy and Schadenfreude 110

Startle and Surprise 110
Neuronal Circuits for Emotions 110
Limbic Circuit 111
Insular Cortex 112
Cingulate Cortex 113
Prefrontal Cortices 113
Amygdala 114
The Bed Nucleus of the Stria Terminalis (BNST) 115
Basal Ganglia 115
Thalamus 117
Hypothalamus 118
Midbrain 119
Hippocampus 120
Summary and Conclusion 121
References 122

9. Neuromodulators 126
Neuromodulator Molecules 126
Dopamine 126
Norepinephrine 128
Serotonin 129
Histamine 130
Acetylcholine 130
Enkephalin, Dynorphin and Endorphin 131
Oxytocin 131
Cannabinoids 132
Steroids 132
Nitric Oxide 133
Drugs of Abuse and Addiction 133
Physical Neuromodulation 134
Summary and Conclusion 135
References 137

10. Sleep and Dream 139
Evolutional Origin and Purposes of Sleep 139
Necessity of Sleep and Variation of Sleep Time 140
Circadian Rhythm 141
Levels of Sleep and Brain Waves (Electroencephalogram) 141
REM Sleep 143
Neuronal Mechanisms of Sleep Regulation 143
Different Types of Dreams and Their Roles 144
Neural Correlates of Dreams 145
Self and Consciousness in Dreams 146
Summary and Conclusion 148
References 150

11. Psychiatric and Neurodevelopmental Disorders **151**
Brain Disorders 151
Schizophrenia 152
Dissociative Disorders 154
Neurodevelopmental Disorders 155
Autism Spectrum Disorders 155
Attention Deficit Hyperactivity Disorders 156
Genetic Variations Related to Schizophrenia and Autism Spectrum Disorders 157
Depression 157
Anxiety Disorders 158
Bipolar Disorders 158
Post-Traumatic Stress Disorders 159
Psychopathy 160
Epilepsy 160
Excitation and Inhibition Balance 161
Dementia 162
Alzheimer's Disease 162
Parkinson's Disease 163
Huntington's Disease 164
Prion Disease 164
Summary and Conclusion 165
References 166

12. Short-Term and Long-Term Merits and Demerits for Individuals, Groups and Species **169**
Conflicts of Short-Term and Long-Term Merits and Demerits 169
Conflicts of Merits and Demerits Between Individuals, Groups and/or Societies 170
Decision Making in Conflicting Conditions and Development of Metacognition 171
Game Theory 171
Roles of Self-Centered Emotions in Decision Making 172
Size of The World and Human Power 173
Past Experience and Education 174
The Emergence of Distinctive Rules in Higher Order Systems 174
Purposes and Means 175
Judgement, Prediction and Decision Based on Facts, Reasoning, Opinion, Belief, Hope and/or Emotions 176
Facts 176
Various Ways of Reasoning 177
Opinion, Belief, Hope and Emotions 177
How We Should Judge and Decide 178
Summary and Conclusion 180
References 181

13. Reconsideration of Self, Mind and Consciousness **182**
What is The Brain for, and What is Consciousness? 182
Consciousness of AI 184
Self, Mind and Consciousness in Different States 186
Objective Centers of Metacognition 187
Unity of Consciousness and Self 188
Summary and Conclusion 189
References 190

14. Concluding Remarks **191**
Aims of This Book 191
Main Conclusions 192
Critical Remarks on Main Conclusions 194
Unresolved Questions and Future Problems 195
Final Message to Readers 198

Index **199**

1

Self, Mind and Consciousness

What is Myself?

I guess that most of you have had the opportunity to think about yourself. At a turning point of your life, you might have considered what you really wished for. Before a job interview, you might have thought about your merits and demerits, laying emphasis on your strong points. However, the fact is that most people do not always know themselves very well. There might be some characteristics and/or habits that you are not aware of, and you might have recognized them only after they were pointed out to you by others. I, myself, had such experiences, that strong emotions came forth at unexpected times, upsetting me greatly. "Myself" is mysterious.

"What is myself?" has been my long-standing concern. By the way, both "myself" and "mind" refer to subjective existences, and "consciousness" refers to subjective states and/or subjective experiences. What are their definitions and relations? According to Cambridge Learner's Dictionary, myself is used to emphasize "I", and self is defined as "the set of someone's characteristics, such as personality and ability that are not physical and make that person different from other people". Self seems to be closely related to mind. Then, what is mind? It is explained by the dictionary as "the part of a person that makes it possible for him or her to think, feel emotions, and understand things". This definition can be interpreted to mean that mind is the main body making a person conscious, because consciousness is defined as "the state of understanding and realizing something" or "the state of being awake, thinking, and knowing what is happening around you".

A philosopher, René Descartes's famous phrase, "I think, therefore I am" has been interpreted to convey that nothing but the existence of 'myself', who is thinking and feeling is without any doubt; implying that other existences in the environment or the world can be questioned. Certainly, I recognize my family, friends, other people, societies, nations, nature and the world by perceiving and recognizing them through my sensory organs and nervous system, as explained in the following chapters. Therefore, if I lose all of them, all the present world will disappear for myself. Actually, we experience loss of consciousness and the universe vanishes every day, when we fall asleep (Tononi et al. 2024). To me the world exists as I can perceive it directly or indirectly by some means. Thus, each person's

DOI: 10.1201/9781003518822-1

self is an indispensable existence. Through this book, I would like to explain the physicochemical or neuroscientific bases and backgrounds of the mysterious self, mind and consciousness, and to discuss how they have developed evolutionally to merit animals and humans.

Definitions of Consciousness

Firstly, I would like to define "consciousness" further, because the word is used in different ways and has multiple meanings (Zeeman 2001, Morin 2006, Bayne et al. 2016, Frith 2021). One meaning is related to the state or level of wakefulness. The second meaning is awareness or being aware of something around you. The third meaning is deep understanding and thinking of the environment and yourself, which is closely related to the Descartes's "I". They are related but different from one another, and it is important which "consciousness" is referred to at each occasion. In the following sections, I explain them further.

Wakefulness

In some cases, consciousness can be equal to wakefulness. Medical doctors judge the level of the wakefulness of a patient by using a predetermined criteria such as the Glasgow coma scale, in which, eye opening, verbal response and motor response are evaluated (Teasdale and Jennett 1974). As for eye opening, four points are given when a patient opens eyes spontaneously or by a usual call, three points when a patient does so after a loud call, two points when a patient does so if painful stimulation is applied, and one point when a patient does not open the eyes even if painful stimulation is applied.

As for the verbal responses, five points are given when a patient speaks retaining "orientation", which is a general knowledge or understanding of oneself and one's situations in the environment such as one's name, age, date and where one is. Four points are given when a patient has a conversation with disturbed orientation, three points when a patient speaks inappropriate words, two points when a patient says something meaningless, and one point if a patient says nothing.

As for motor responses, six points are given when a patient moves their limbs as instructed, five points when a patient brushes away a painful stimulus by a hand, four points when a patient withdraws a limb if a painful stimulation is applied to it, three points when a patient shows slow flexion movement when receiving a painful stimulation, two points when a patient shows slow extension movement when receiving a painful stimulation, and one point when a patient does not move. The level of wakefulness is judged higher with a higher score, and the lowest score—three points corresponds to deep coma. Thus, wakefulness or this type of consciousness can be recognized objectively.

Awareness

In many cases "being conscious" means "being aware of something". We sense external information through our sensory organs such as the eyes, the ears, the nose, the tongue and the skin. Such information is processed by our nervous system including the brain; the information is perceived and recognized. Recognized objects are often considered further, and utilized in action selection and/or remembered. It is to be noted that we are not aware of all the external information reaching our sensory organs; and only a small portion is recognized. In a crowed place, you may concentrate on what your friend is saying and may not perceive other noises. Similarly, you usually do not perceive all visual signals coming to your eyes. Only interested or conspicuous signals or objects are selected, perceived and recognized. Attention facilitates the selection process of input signals, and uninterested information is not recognized (Walther and Koch 2007, Moore and Zirnsak 2017). Sensory information processing is explained further in chapter four, and attention is explained in chapter five.

However, it is to be noted that unnoticed external information may influence the behavior or action of a person. Reflexive movement occurs unconsciously by receiving a sensory stimulus. When a limb muscle is stretched, it contracts by stretch reflex mediated through the spinal cord. When your head or body is rotated, your eyeballs turn in the opposite direction in order to keep the visual image on the retina stationarily, which is called vestibulo-ocular reflex, and mediated by semicircular canals in the inner ears and the brainstem neuronal pathways (Robinson 2022). Motor control mechanisms are explained further in chapter four. A primary or original role of the nervous system is to transmit external signals detected by a sensory organ to the motor systems, so that an animal can respond appropriately to the stimulation that has reached the animal. The brain seems to have developed in order to improve the animal's motor or behavioral responses to stimuli through sophistication of recognizing objects, understanding the situation, and the learning and decision-making processes.

It is also known that in some cases unnoticed or unrecognized sensory information affects the complicated behavior of a person. If a few images, showing a bottle of drink are inserted into a movie, most viewers do not notice the image of the bottle, but some become thirsty and feel like having the drink. Such a phenomenon is called subliminal perception, and suggests that unnoticed or unrecognized information could affect one's behavior or decision making (Sandberg et al. 2022). The blindsight is another example where an unaware visual stimulus affects the behavior of a person. For example, a patient with a defect in the primary visual cortex claims not to see a visual stimulus at all, but he or she can detect and localize such an unseen stimulus (Stoerig and Cowey 2007).

Qualia

When we perceive and recognize something, we have subjective experiences. The subjective feeling or emotion may be a nostalgic scenery, calming or inspiring music, delicious smells, freezing coldness or biting pain. Such a vivid feeling is called qualia (Koch 2004, Kanai and Tsuchiya 2012). Qualia are caused by neuronal activities in our brain, but how biophysical or biochemical mechanisms in our neuronal systems give rise to subjective feelings is a very difficult question to answer. Because qualia are subjective, we can only guess at other people's qualia through introspective observation or thinking of our own qualia. We know the color red is accompanied with certain qualia, and usually think that other people's qualia to red are the same or similar. However, strictly speaking, there seems to be no way of verifying whether other people's qualia are like yours.

Then, how we can say that subjective qualia are caused or related to our sensory systems and/or neuronal activities in the brain? There are color blind people, and among them some have difficulty in distinguishing the colors red and green (Moussawi et al. 2021). The qualia related to a color of a color-blind person must be different from others, which implies that qualia depend on the characteristics of our physicochemical sensory system: a protein in photo-receptor cells in this case. Qualia cannot stand by themselves, independent of our sensory and nervous systems.

Visual agnosia is a class of clinical phenomena defined as impaired object recognition that cannot be attributed to visual loss, language impairment, or a general mental decline (Biran and Coslett 2003). Some patients seem to generate an internal representation of the stimulus, but fail to recognize it. For example, when a patient sees an apple and is asked to draw its picture, he or she draws an inappropriate or meaningless picture. However, when asked to draw a picture of an apple without seeing it, the patient apparently can do it. In this case, the concept of an apple is retained in the patient, but he or she fails to recognize an apple by seeing it. The patient's qualia generated by visual inputs seem to be different from others, and indicates that the qualia of a particular person are distinct and owned only by himself or herself. The existence of such disturbance of perception and/or recognition in some people suggests that qualia depend on the physicochemical machinery of a particular person.

It is also to be noted that one's qualia may be distorted by specific neuronal mechanisms that are explained in chapters five and eight. Illusion provides qualia that are somewhat different from that provided by a real object (Todorović 2020).

Self-Recognition and Metacognition

The highest level of consciousness is related to deep understanding and thinking of recognized and memorized objects in our brain, including information about ourselves, for which recognition of self is required (Morin 2006, Fleming 2024). Descartes's "I" is based on this type of consciousness. The function of searching and processing conceptualized materials and information in ourselves is called metacognition. I think that the primary role of metacognition is improvement of decision-making and action selection, utilizing many related information which have been obtained and retained by an individual.

The appropriate selection of behavior should merit an individual by increasing the probability of getting food or a partner of the opposite sex, by improving the individual's status within the group, and/or by decreasing demerits or risks. Nobody can be certain about which is the best or better choice of behaviors in most cases, because there are usually some conflicts between merits and demerits, and between short-term and long-term effects. Metacognition enables evaluation and comparison of each one of them.

Self-recognition clarifies and highlights oneself as distinct from others, and helps to consider the merits and demerits of oneself. It also seems to facilitate speculation of other persons' thoughts . The ability to guess what is in the other person's mind is called "theory of mind", which is explained in the next section.

How self-recognition has developed evolutionarily is an interesting question. Self-recognition has been tested by observing an animal facing a mirror, which has suggested that not only chimpanzees and dolphins, but also birds and fish, show some behaviors which could be interpreted to rely on self-recognition, although the levels of self-recognition seem to vary among animal species (Lage et al. 2022).

Theory of Mind

The theory of mind is the ability to suppose another individual's mind. Here I explain it referring a scheme shown in Figure 1. Two people A and B are in a room in which there are a bookshelf, a table and chairs. Person A sees that B is sitting on a chair

Figure 1. Explanation of the theory of mind. Person A sees person B in a room. (1) B is reading a book sitting on a chair in front of a table. After a while B hears a phone call from outside of the room, and B goes out from the room leaving the book on the table. (2) Then, another person C comes into the room and finds the book on the table. (3) C brings the book to a bookshelf and goes out from the room. (4) B comes back and finds no book on the table. At this timing, A considers that B may be wondering what has happened to the book which B was reading before B left the room (Hirano 2019).

and reading a book in front of the table. Then, B hears a phone-call from outside of a room, puts the book on the table and gets out from the room. After that, another person C comes into the room, finds the book on the table and puts it on the bookshelf, and leaves the room. After a while, B comes back to the room and finds no book on the table. At this moment A considers that B is probably wondering what has happen to the book that he was reading before he left the room. Such speculation by A about B's mind and thoughts, is an example of the theory of mind.

Self-recognition seems to be required for the theory of mind to function. In the theory of mind, one considers the other's mind by replacing the other's condition or state with his or hers. The theory of mind is likely to contribute towards improved communication between individuals and coordinated behaviors of members of a group. Such improvement seems to make competitions between individuals in a group advantageous. However, it is to be noted that the speculation using the theory of mind is based on one's mind and one's own personal experiences. The experiences and the mind of individuals can be largely different, and in such cases the supposition of another's mind can be inappropriate or misunderstood. It has been reported that the theory of mind develops to a certain level, that is understanding of other's belief and the reality, between the ages of three and five (Poulin-Dubois 2020).

Hard Problem

To link subjective qualia, consciousness and mind with objective biophysical and biochemical processes in the nervous system or the brain is very difficult, if not impossible. Thus, this question has been called the hard problem, as was named by a philosopher David Chalmers (Charmers 2023). He also developed the idea of hypothetical existence of a philosophical zombie, who is indistinguishable from a human except that he or she has no consciousness or mind. The concept of a philosophical zombie has been used for philosophical discussion about the subjective mind (Fischer and Sytsma 2021). Strictly speaking, it is impossible for us to say that other people are not philosophical zombies. Because consciousness is subjective and belongs to only one individual, therefore, others can not directly address it. Conversely, if an artificial intelligence (AI) responds to any questions in a manner that is indistinguishable from the response of humans, it cannot be denied that it has consciousness in principle. This issue is discussed again in chapters seven and thirteen.

Thus, there is a huge gap between the subjective mind and consciousness, and objective biophysical and biochemical phenomena in the brain and sensory organs. However, as I have explained, we know that physical or chemical factors in a human affect his or her qualia. Thus, consciousness and mind must be brought about or based on objective biological materials in our body. I think that trials to explain completely subjective phenomena by objective materials may not be appropriate or wise. However, it might be possible to clarify objective physicochemical phenomena related to or underlying subjective phenomena about consciousness and mind, and to discuss how and why such subjective phenomena have emerged.

Neuronal Correlates of Consciousness (NCC)

Attempts to determine the neuroscientific bases of consciousness have been made. In particular, awareness has been intensively studied. Some researchers have tried to identify neural correlates of consciousness (NCC), which has been defined as the minimum neuronal mechanism jointly sufficient for any one specific percept (Koch et al. 2016).

An experiment to identify brain regions related to the perception of a face using functional magnetic resonance imaging (fMRI) is schematically explained in Figure 2. A research participant sees an image of a face with visual noises, while fMRI is performed. The participant may or may not be aware of the face. The fMRI image is recorded in both cases, and the latter fMRI image is subtracted from the former. The subtracted fMRI image points out brain areas related to the recognition of a face, although it is possible that some of the areas are involved in functions other than the recognition of a face itself, such as selective attention and unconscious stimulus processing (Koch et al. 2016, Nani et al. 2019).

NCC could be identified in more detail. Particular activities of specific neurons could be NCC for percept of a face. In this case, if such neurons are somewhat activated artificially in a person, he or she should see a face even if it is not present.

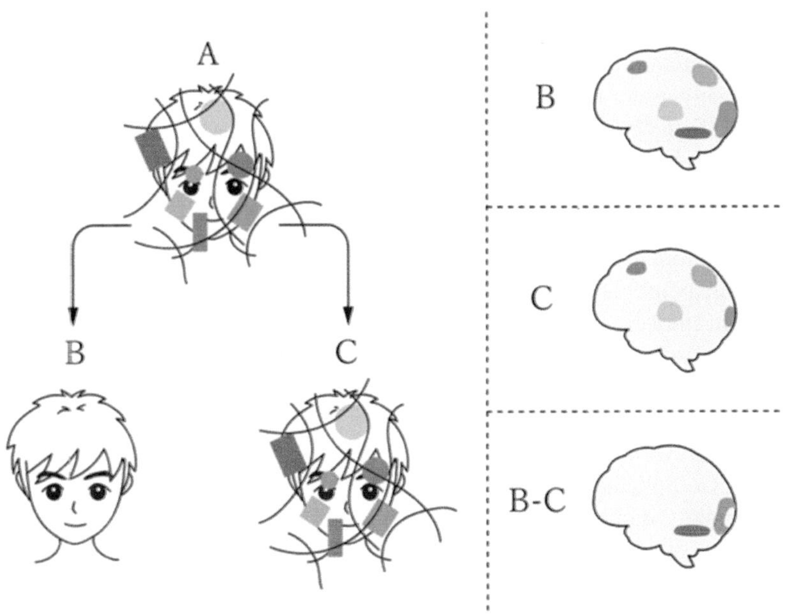

Figure 2. Schematic explanation of an attempt to identify regions related to perception of a face. An image of a face with visual noise (A) is presented to a research participant and fMRI is performed. The participant may (left B) or may not (left C) recognize the face. fMRI image is recorded in both cases (right B and C), and the latter is subtracted from the former. The subtracted fMRI image (B–C) shows the brain regions related to perception and/or recognition of the face, although regions not directly related to perception and recognition could be included. Read the text (Koch et al. 2016, Hirano 2021).

On the other hand, if these neuronal activities are suppressed, one could not perceive a face even if there is a face (Koch et al. 2016). Thus, certain activities of specific neuronal populations can be proposed to underly subjective perception or awareness. However, I would like to note several points about the identification of NCC. Firstly, activation of only a certain portion of whole neurons related to the perception of an object could be sufficient to cause perception or generation of qualia of an object, as the neuronal activities in a certain portion might be able to activate related neurons outside of the portion. A proposed NCC that is identified as a neuronal population capable of causing a percept when stimulated, could be smaller than the whole neuronal population that is responsible for the percept, and multiple partial populations might be identified as NCC. Secondly, there may be also redundancy in neurons related to one percept which suggests that specific neurons cannot be definitely identified as NCC. I also would like to note that some neurons can be multifunctional: they could contribute to multiple qualia and possibly to functions other than perception or recognition. Furthermore, activities of each neuron are quite variable and may be different in each case. I wonder whether qualia related to a single object may be different to a certain extent in each occasion. More detailed neuronal mechanisms underlying perception, recognition and awareness are explained and discussed in chapters four and five.

Unconscious Processes

This book primarily focuses on the mind and consciousness. However, I would like to note that many motor responses to stimuli occur unconsciously. As explained above, reflexes mediated by the spinal cord or the medulla oblongata, such as stretch reflex or vestibulo-ocular reflex take place automatically, and we are usually unaware of them. Autonomic responses such as secretion of saliva during eating, digestion and nutrient absorption in digestive organs, and control of heart beat rate and breathing also occur automatically. We do not pay attention to them, in spite of the fact that they are indispensable to life and are primarily controlled by our nervous system, as explained in chapter four. Actually, the nervous system seems to have been developed to convey information from the sensory organs such as the eyes and the ears to motor systems such as the muscles and secretory organs.

Rather complicated motor actions can be performed unconsciously. For example, when we walk, we do not think about how to move each leg. We just decide where we wish to go. Complicated limb movements during walking are controlled by particular neuronal circuits named the "central pattern generator" which is explained in chapter four (Minassian et al. 2017). Thus, many motor responses of humans and animals which seem purposeful can take place without the participation of the consciousness or the mind.

Even a one cell organism such as a *Paramecium* shows behaviors which seems purposeful (Brette 2021). It has molecules to sense external stimuli such as touch, light, nutrient and temperature, and also motor systems such as cilia. When a *Paramecium*'s forward movement is prevented by an obstacle such as a stone, it moves back by reversing the ciliary movement, changes direction and moves forward again. The speed of a *Paramecium*'s forward movement is accelerated when it is touched at the rear end; which has been interpreted to mean that the *Paramecium* tries to escape from a predator. A *Paramecium* avoids bright light, and it also tends to move toward the water surface. Thus, a *Paramecium* shows rich varieties of behaviors which seem purposeful in spite of the fact that it is a unicellular organism. Plants without nervous systems also show purposeful movements. For example, carnivorous plants can catch insects and utilize them as nutrients (Ellison and Gotelli 2001).

Multicellular animals including vertebrates show complicated but prefixed behaviors. One example is filial imprinting (McCabe 2019). A precocial bird such as a duck, chicken or quail establishes a strong social attachment with the first seen moving object, which is in most cases, a parent. The young bird learns the characteristics of the object and approaches it. Such behavior helps the survival of the young birds.

Most of you would not think that a *Paramecium* or a plant has consciousness and/or mind. Then, what about a young bird? Complicated and apparently purposeful behaviors might take place possibly without the participation of consciousness. Thus, the existence of consciousness in organisms other than ourselves is questionable, and in what evolutional and developmental stages consciousness and mind emerge are interesting but difficult questions, which are discussed further in chapter two.

Summary and Conclusion

Self, mind and consciousness are subjective existences and phenomena. They are based on objective biophysical and biochemical materials and rules. Attempts to correlate neuronal activities to consciousness have been carried out. Consciousness has multiple meanings such as an arousal state, awareness and metacognition including self-recognition. When we are aware of something, we have vivid personal feelings called qualia. The neuroscientific bases of qualia have been studied as the search for neural correlates of consciousness (NCC), which is defined as the minimum neuronal mechanisms jointly sufficient for any one specific percept. However, rigorous identification of NCC may be difficult, as there may be redundancy in neurons related to one percept and the activities of each neuron may be different in each situation. Metacognition is the highest level of consciousness and would improve the quality of decision-making, utilizing many related information including memory. Self-recognition would help to consider short-term and long-term merits and demerits for oneself, and to think about the selves of other persons and what they have in their minds. A direct explanation of the subjective mind and consciousness with objective biophysical and biochemical neuronal activities, is a hard problem to be resolved. This is because the mind and consciousness are personal. Strictly speaking we can only guess what is in another person's mind and consciousness through the introspective observation and thinking. It is also to be noted that the motor responses of many humans or animals to external stimuli occur unconsciously, and that apparently purposeful behaviors can take place without the participation of consciousness.

References

Bayne, T., J. Hohwy and A.M. Owen. 2016. Are there levels of consciousness? Cogn. Sci. 20: 405–413.
Biran, I. and H.B. Coslett. 2003. Visual agnosia. Curr. Neurol. Neurosci. Rep. 6: 508–512.
Brette, R. 2021. Integrative neuroscience of *Paramecium*, a "swimming neuron". eNeuro 8: 18–21.
Cambridge Learner's Dictionary. 4th Ed. 2012. Cambridge Univ. Press. Cambridge, U.K.
Charmers, D.J. 2023. David, J. Charmers. Neuron 111: 3341–3343.
Ellison, A.M. and N.J. Gotelli. 2001. Evolutionary ecology of carnivorous plants. Trends Ecol. Evol. 16: 623–629.
Fischer, E. and J. Sytsma. 2021. Zombie intuitions. Cogn. 215: 104807.
Fleming, S.M. 2024. Metacognition and confidence: a review and synthesis. Annu. Rev. Psychol. 75: 241–268.
Frith, C.D. 2021. The neural basis of consciousness. Psychol. Med. 51: 550–562.
Hirano, T. 2019. What is the Brain for? Behavioral Selection and Neuroscience in the Era of AI (Japanese) Kyoto University Press, Kyoto, Japan.
Hirano, T. 2021. What is Myself? Neuroscientific Consideration of Self and Mind (Japanese) Gentousya Press, Tokyo.
Kanai, R. and N. Tsuchiya. 2012. Qualia. Curr. Biol. 22: R396.
Koch, C. 2004. Qualia. Curr. Biol. 14: R496.
Koch, C., M. Massimini, M. Boly and G. Tononi. 2016. Neural correlates of consciousness: progress and problems. Nat. Rev. Neurosci. 17: 307–321.
Lage, C.A., D.W. Wolmarans and D.C. Mograbi. 2022. An evolutionary view of self-awareness. Behav. Processes 194: 104543.
McCabe, B.J. 2019. Visual imprinting in birds: behavior, models, and neural mechanisms. Front. Physiol. 10: 658.
Minassian, K., U.S. Hofstoetter, F. Dzeladini, P.A. Guertin and A. Ijspeert. 2017. The human central pattern generator for locomotion: does it exist and contribute to walking? Neuroscientist 23: 649–663.
Moore, T. and M. Zirnsak. 2017. Neural mechanisms of selective visual attention. Annu. Rev. Psychol. 68: 47–72.
Morin, A. 2006. Levels of consciousness and self-awareness: A comparison and integration of various neurocognitive views. Conscious. Cogn. 15: 358–371.
Moussawi, Z.E., B. Marguerita and Al-Haddad. 2021. Gene therapy in color vision deficiency: a review. Int. Ophthalmol. 41: 1917–1927.
Nani, A., J. Manuello, L. Mancuso, D. Liloia, T. Costa and F. Cauda. 2019. The neural correlates of consciousness and attention: two sister processes of the brain. Front. Neurosci. 13: 1169.
Poulin-Dubois, D. 2020. Theory of mind development: state of the science and future directions. Prog. Brain Res. 254: 141–166.
Robinson, D.A. 2022. Basic framework of the vestibulo-ocular reflex. Prog. Brain Res. 267: 131–153.
Sandberg, K., S.H.D. Pin, M. Overgaard and B.M. Bibby. 2022. A window of subliminal perception. Behav. Brain Res. 426: 113842.
Stoerig, P. and A. Cowey. 2007. Blindsight. Curr. Biol. 17: R822–824.
Teasdale, G. and B. Jennett. 1974. Assessment of coma and impaired consciousness. A practical scale. Lancet 13: 81–84.
Todorović, D. 2020. What are visual illusions? Perception 49: 1128–1199.
Tononi, G., M. Boly and C. Cirelli. 2024. Consciousness and sleep. Neuron 112: 1568–1594.
Walther, D.W. and C. Koch. 2007. Attention in hierarchical models of object recognition. Prog. Brain Res. 165: 57–78.
Zeeman, A. 2001. Consciousness. Brain 124: 1263–1289.

2

The Purpose of Life and the Role of the Brain

What is Life?

Humans as well as other animals, plants and bacteria are lives or living creatures. The most important characteristics of lives is that they prolificate: they produce offsprings which are similar to themselves. A group of lives having similar characteristics and proliferating by mating within the group is called a species, although some lives such as bacteria and yeast can proliferate by cell division without mating. A large number of animal, plant and bacteria species have emerged through evolution. All lives on the earth are primarily composed of organic macro-molecules, which are proteins, nucleic acids, lipids and carbohydrates (Alberts et al. 2023).

Whether viruses are lives or not has been discussed (Piast 2019), as they can proliferate only in cells of other living creatures and can be crystalized just like inorganic or organic molecules such as proteins. Viruses could be regarded as lives as they proliferate. However, they can proliferate only in specific cells of other organisms, as their reproductive machineries are incomplete, and outside of such a cell, viruses are inactive just like inanimate materials. Thus, the inclusion of viruses to lives has been questioned. I think that whether viruses are lives or not is not important, because the definition of life depends on what people choose to believe in. In any case, it should be noted that the primary purpose of life seems to be its long-lasting survival as a species.

The Survival of an Individual and a Species

The main purpose of a species as a whole seems to be its survival as a species and the growth of its population. On the other hand, the main purpose of an individual life seems to be its survival and the generation of its own offsprings in general. The purposes of individuals are different among various species.

We humans usually care about ourselves, families, friends and/or the relatively small groups that we belong to more than the merits of a large society; such as a company, a nation or all human beings. Of course, many people are worried about

DOI: 10.1201/9781003518822-2

world-wide problems such as global warming, nuclear weapons, etc. Nevertheless, the merits and/or demerits of an individual and/or those of a relatively small group of people matter a lot for most people. I think that this is so mainly because our emotions are self-centered and they are induced by our own experiences as explained further in chapter eight.

Worker bees or ants behave as if the merits of their colony are more important than their own merits or lives (Luo 2021, Chittka and Rossi 2022). In their nest only one queen gives birth to all members of the colony, and there are hundreds or thousands of workers. The workers are all females, but they do not give birth to any offsprings. They gather food for the colony members, take care of the eggs and larvae, and protect the nest.

I do not think that an individual in any animal species behaves or considers contributing to the species as a whole directly. Rather, an individual animal tries to survive, gives rise to its own offsprings and/or contributes to a relatively small group, to which it belongs. Such efforts of an individual animal would eventually support the survival and prosperity of the species in most cases.

Various Ways of Keeping a Species

There are different ways of keeping a species. One strategy is to give rise to many offsprings, and the other is to raise a small number of offsprings, possessing high abilities for the survival. Some fish lay a huge number of eggs. For example, it is said that a sunfish lays about three hundred million eggs at a time. However, the survival rate of sunfish larva seems to be very low, and the number of sunfish is kept within a certain range.

On the other hand, the number of offsprings of a mammal species is relatively small. Humans or other large mammals have one or a few babies at a time, and their survival probabilities are relatively high. Among them too, different strategies are adopted for the survival. A young human takes a long time to grow up, during which time he or she learns a lot and obtains high intelligence and adaptive abilities that support the survival. On the other hand, herbivores such as horses and cows are born as relatively matured animals. They stand up immediately after birth, presumably so that they can run away from predators.

The importance of a juvenile individual's life in the survival of the species as a whole, seems to be different among species; although the number of individuals in a species is usually kept within a certain range.

Self, Offsprings and Groups

The relative importance of the self, its offsprings and the group it belongs to varies among different species and also among different individuals in an animal species. Amongst humans, some individuals may give the first priority to their own merits and demerits. For others, the merits of their children may be more important. Some others may preferentially choose behaviors that contribute to the group they belong to, such as a family, a company, a religious group or a nation.

The aims of behaviors that merit an individual are multiple, and some help in their survival. Among these behaviors are searching for and finding food and escaping from a predator or an enemy. Finding a partner for reproduction and getting a higher position within a group also merit an individual. Which of these merits is chosen by an individual depends on the situation and characteristics of the individual. If an animal finds that a predator is approaching, escape from it will be given a high priority. However, there are cases where higher priority is given to the merits of offsprings or a group than the life of an individual.

Some fish, such as salmon, expend tremendous effort to reach a breeding spot located upstream in a river (Luo 2021). There they mate, breed, get exhausted and die. Such behaviors appear to give higher priority to breeding than to the life of the individual. Thus, giving rise to offsprings can be more important than an individual life in some animal species, especially during specific periods of their lives. As described previously, the behavior of worker bees is primarily for the colony. Worker honey bees fight with wasps that are attacking their nest, even though the wasp is much larger and stronger than the bee. This behavior of the bee shows that it gives higher priority to protecting the group than to its own safety. Although aims of animal behavior are diverse, in principle, all of them ultimately contribute to the survival of the species.

How animals recognize or identify members of con-species is an interesting question. There seem to be both inborn and learning-dependent mechanisms. As explained in chapter one, a precocial duck shows a strong attachment to the first seen moving object. Songs contribute to the recognition of con-species in some birds (Louder et al. 2019), and chemical substances such as pheromones contribute to con-species recognition in many animal species (Liberles 2014). In some cases, such as explained in the next section, a mammal can identify an individual con-species through chemical substances.

Self-Interest and Altruism

The behavior of worker bees contributing to a colony seem to be altruistic. There are many other examples of altruism (Wyatt 2017). A monkey helps another monkey so that the latter can obtain food (Burkart et al. 2014). Altruistic behavior has also been reported in rats (Hernandez-Lallement et al. 2020). Rats avoid actions that harm other rats. A previous pain experience of a rat enhances behavior selection not to harm another rat, although the harm aversion is suppressed by a larger merit to the rat. If a hungry rat can get a large amount of food by taking an action that results in harming another rat, the former rat chooses the behavior that harms another. It seems that rats have a kind of theory of mind, and can also perform metacognitive process to compare different merits for decision making.

Humans collaborate with other animal species such as dogs, horses etc. In many cases humans and dogs behave to merit each other. In some cases, one helps the other even if there is a risk of injury. Animal species other than humans also collaborate with other animal species (Wyatt 2017). Some coral reef fishes such as coral trout indicate a hidden prey to cooperative hunting partners such as giant moray eels or octopuses (Vail et al. 2013). In this case, both the collaborating species obtain merits in hunting.

On the other hand, behaviors to harm or kill other conspecifics have also been reported. A male of a monkey species, Hanuman Langur, practices infanticide, presumably in order to increase its own babies (Ostner et al. 2006). There is another mechanism of increasing a male's offsprings which is known as the Bruce effect (Bruce 1959). It is a kind of pregnancy block or suppression that occurs in a female rodent following exposure to an unfamiliar male. Bruce effect is not a behavior, but a phenomenon contributing to increasing the number of offsprings of a particular male. Thus, there are behaviors and other mechanisms that contribute to increasing the number of offsprings of a particular male animal at the sacrifice of conspecifics' lives. By the way, Bruce effect is mediated by chemical substances, and is an example of recognizing an individual conspecific through chemical substances (Liberles 2014).

The idea that genes work to reproduce themselves using the body of living creatures, and that animals' behaviors are determined by the efficiency of the contribution to gene reproduction, has been proposed (Dawkins 2006). According to this theory, the behavior of worker bees contributing to the colony can be interpreted to mean that they contribute to keeping certain amounts of their own gene, as parts of worker bees' gene are same as the queen's gene (Burke 2014). Thus, apparently altruistic behaviors might bring some merits to the individual such as the contribution to keep some parts of its own gene, although some altruistic behaviors such as rat's behavior to avoid harming another rat do not seem to provide clear and immediate merits to the animal.

Recognition of Environment and Selection of Action

In principle, I think that behaviors of all animals should ultimately contribute to the survival of the species. Behaviors for the survival of an individual and those for reproduction should be useful for the survival of the species. The individual's competitive behaviors in a group might contribute to the selection of a good leader possessing a strong body and/or a high adaptive ability—characteristics that are likely to support the survival and/or prosperity of the group.

An animal chooses behaviors by recognizing the state of its environment and itself. Physicochemical signals from the environment such as light, sound, touch and chemical substances and also those from its own body such as a low glucose level in the blood trigger actions as explained in chapter four. There are different types of motor actions or behaviors primarily based on different levels of perception and recognition. Reflexes are triggered by simple stimuli, and some voluntary actions are initiated after metacognitive processing of various recognized and memorized information. All of these behaviors seem to have developed to help the survival of the species during evolution. Different types of behaviors are explained further in the next section.

Reflex, Instinctive and Voluntary Actions

Reflexes such as stretch reflex, vestibulo-ocular reflex or eyeblink on receiving a puff of air occur unconsciously. If a sound application that occurs before the puff of air is repeated, eyeblink comes to take place by the sound application alone (Christian and Thompson 2003). Such a phenomena is an example of conditioned reflex, which also occurs unconsciously. It is usually difficult, if not impossible, to suppress any reflexes, including conditioned reflexes. In both unconditioned and conditioned reflexes, the trigger stimulus is usually simple, such as the stretch of a muscle, a body turn, or a mechanical stimulation to a particular part of the body or a simple sound.

Instinctive behaviors are composed of actions of a relatively fixed pattern, which are triggered by key stimuli called releasers, and are considered innate in principle; although recent studies have suggested that some of them develop under the guidance of species-specific experiences (Blumberg 2017, Luo 2021). Examples of such behaviors include the migration of birds and the reproductive behaviors of fruit fly or fish (Luo 2021, Oka 2023). The reproductive behavior of fish is influenced by hormone, pheromone, and/or the behavior of an individual of the opposite sex, etc. This shows that the triggering stimuli for instinctive behaviors and induced actions are more complicated than those of simple reflexes. When an appropriate stimulus is applied in an adequate condition, it may be hard for an animal not to behave in an instinctive manner. In this sense, instinctive behavior may be somewhat automatic and/or compulsory. Whether an animal experiences any subjective feelings to a releaser stimulus of instinctive behavior, and if so, what kind of qualia it has are interesting questions that are difficult to answer.

We humans often choose an action from a number of possibilities. Such a choice of behavior is not compulsory in principle; metacognition plays an important role in voluntary behavior. The choice of voluntary behavior is largely influenced by the situation of an individual and his or her environment, and also by the characteristics and remembered experiences of the individual. We could guess to a certain extent what kind of voluntary behavior a particular individual would exhibit using the theory of mind. Thus, the choice of the voluntary action might not be completely unexpected. Animals other than humans such as dogs, cats, birds, etc. also appear to show voluntary actions. To make a distinction between a voluntary behavior and a highly complicated instinctive behavior might not be so easy.

Emergence of The Consciousness and The Mind

At which stage of evolution do animals obtain mind and consciousness? In other words which animal species have mind and consciousness? This is a difficult question. The answer seems to depend on definitions of consciousness and mind. If consciousness means an objective arousal state or wakefulness, most animals seem to have it. Objective awareness also appears to be possessed by most animals, because they try to escape from a predator, when they find and recognize it. Objective metacognitive ability including the theory of mind seems to be present in some mammals. As described in a previous section, a rat avoids a behavior that harms another rat.

The problem is whether they have subjective consciousness, emotions and feelings such as we humans do. Strictly speaking, it is impossible to answer this question, because we cannot directly experience the consciousness of other people or animals.

Key points might be whether the choice of behavior or motor action is automatic or voluntary, and/or whether there are conflicts between the merits and demerits or short-term and long-term effects, because we are usually unaware of or do not remember reflexive or automatic actions.

A unicellular organism *Paramecium* appears to find a way to procced and also try to escape from a predator as if it has its own will (Brette 2021). However, most people would not think or feel that *Paramecium* without a brain or a nervous system, has mind and consciousness. Actually, the above reactions of a *Paramecium* are triggered by receptor molecules on the cell membrane, and the information is conveyed to motor machinery cilia using functions of membrane proteins and movements of ions, which then changes the direction of ciliary rotation, resulting in altered locomotion of *Paramecium*. *Paramecium*'s behavior could be explained only by physicochemical reactions, and could be regarded automatic.

Cnidarians such as jellyfish and hydra have a dispersed nervous system without a clear center of the nervous system or a brain. Hydra shows several behaviors controlled by a relatively small number of neurons (Yuste 2024). The limited behavioral repertories with fixed patterns and lack of a brain make me feel that hydra is unlikely to be conscious, although I cannot deny the possibility that subjective consciousness exists in a dispersed neuronal network.

What about vertebrates? Fish and birds show complicated innate and/or instinctive behaviors. Patterns of such behaviors look relatively fixed. If these animals showed only innate behaviors of fixed patterns and occurrences of their behaviors were automatic, they might be regarded as being unconscious. However, recent studies have shown that instinctive behaviors are altered by learning (Blumberg 2017). In addition, they show other learned behaviors. Changing complicated behavior by learning, suggests that experiences or memories are considered in decision making and choice of behaviors, implying that a metacognitive process comparing merits and demerits seems to take place.

As described above, a rat avoids behavior harming another rat. But when a large merit is provided by performing the harmful action, the rat chooses the harming behavior (Hernandez-Lallement 2020). In this case, the rat seems to compare each merit, and chooses an action voluntarily using metacognition. Such a behavioral choice makes me feel that a rat might have mind and consciousness somewhat similar to ours.

Then, why have such mind and consciousness emerged during animal evolution? Two meanings of consciousness, awareness and metacognition, contribute to the appropriate choice of behaviors using and evaluating various information from the animal's environment, the inside of its body and/or memory, and are likely to contribute to highly adaptable behavioral choices. Emotions also seem to support such metacognitive functions, which are explained further in chapter eight. Enhanced metacognitive abilities would support competition within the species and the survival of an individual, and also those of the species among other species during evolution. Thus, the reason for the emergence of objective functions of awareness and metacognition could be explained, although the necessity of subjective consciousness, feelings and/or emotions remain enigmatic.

Some people think that subjective mind and consciousness belong to different worlds not directly related to the objective physical world; this is called dualism (Arshavsky 2006, Searle 2007, Demertzi et al. 2009). Presumably these ideas have been born because of the difficulty to answer the hard problem, religious notions, personal beliefs and/or desires. There is also an idea that subjective experiences are unnecessary or non-functional by-products of objective physicochemical phenomena (Robinson et al. 2015). In any case, we know that subjective awareness and qualia are based on objective materials in our sensory organs and the brain. I wonder whether the emergence of subjective mind and consciousness might have somehow promoted or improved the functions of the objective physicochemical brain. I would like to discuss the nature of self, mind and consciousness again in chapter thirteen after explanation of objective nervous systems, emotions, dreams, psychiatric disorders and computer systems.

Summary and Conclusion

The primary purpose of life is considered to be the survival of the species. For an individual animal its own survival and the generation of its own offsprings are critical in most cases. However, in certain instances, some animals such as worker bees behave as if contribution to the group to which they belong are more important than their own lives. Animals show reflex, innate, automatic and voluntary behaviors. In voluntary behaviors an individual chooses an action. For an appropriate choice of behavior, awareness and metacognition using various information from the environment, the inside of the body and/or memory seem to be useful. Such abilities would contribute to obtaining better outcomes in the competition within the species and for the survival of an individual, and they also seem to have contributed to the survival of the species during evolution. Thus, the reason for the emergence of the objective functions of awareness and metacognition might be explained, although the necessity of subjective mind and consciousness remain enigmatic. I wonder if conflicts in decision making might be related to the emergence of subjective consciousness, as we are usually not aware of or remember automatic responses.

References

Alberts, B., K. Hopkin, A. Johnson, D. Morgan, K. Roberts, P. Walter et al. 2023. Essential Cell Biology. 6th Ed. Norton, New York, NY.

Arshavsky, Y.I. 2006. "Scientific roots" of dualism in neuroscience. Prog. Neurobiol. 79: 190–204.

Blumberg, M.S. 2017. Development evolving: the origins and meanings of instinct. Wiley Interdiscip. Rev. Cogn. Sci. 8: 10.1002.

Brette, R. 2021. Integrative neuroscience of *Paramecium*, a "swimming neuron". eNeuro 8: 18–21.

Bruce, H.A. 1959. An exteroceptive block to pregnancy in the mouse. Nature 184: 105.

Burkart, J.M., O. Allon, F. Amici, C. Fichtel, C. Finkenwirth, A. Heschl et al. 2014. The evolutionary origin of human hyper-cooperation. Nat. Commun. 5: 4747.

Burke, A.F. 2014. Hamilton's rule and the cause of social evolution. Philos. Trans. R. Soc. Lond. B Biol. Sci. 369: 20130362.

Chittka, L. and N. Rossi. 2022. Social cognition in insects. Trends Cogn. Sci. 26: 578–592.

Christian, K.M. and R.F. Thompson. 2003. Neural substrates of eyeblink conditioning: acquisition and retention. Learn. Mem. 10: 427–455.

Dawkins, R. 2006. The Selfish Gene. 30th Anniversary Edition. Oxford University Press. Oxford, U.K.

Demertzi, A., C. Liew, D. Ledoux, M. Bruno, M. Sharpe, S. Laureys et al. 2009. Dualism persists in the science of mind. Ann. N. Y. Acad. Sci. 1157: 1–9.

Hernandez-Lallement, J., A.T. Attah, E. Soyman, C.M. Pinhal, V. Gazzola and C. Keysers. 2020. Harm to others acts as a negative reinforcer in rats. Curr. Biol. 30: 949–961.

Liberles, S.D. 2014. Mammalian pheromones. Annu. Rev. Physiol. 76: 151–175.

Louder, M.I.M., S. Lawson, K.S. Lynch, C.N. Balakrishnan and M.E. Hauber. 2019. Neural mechanisms of auditory species recognition in birds. Biol. Rev. 94: 1619–1635.

Luo, L. 2021. Principles of Neurobiology. 2nd Ed. CRC Press, Boca Raton, FL.

Oka, Y. 2023. Neural control of sexual behavior in fish. Zool. Sci. 40: 128–140.

Ostner, J., M.K. Chalise, A. Koenig, K. Launhardt, J. Nikolei, D. Podzuweit et al. 2006. What Hanuman langur males know about female reproductive status. Am. J. Primatol. 68: 701–712.

Piast, R.W. 2019. Shannon's information, Bernal's biopoiesis and Bernoulli distribution as pillars for building a definition of life. J. Theor. Biol. 470: 101–107.

Robinson, Z., C.J. Maley and G. Piccinini. 2015. Is consciousness a spandrel? J. American Philosophical Assoc. 1: 365–383.

Searle, J.R. 2007. Dualism revisited. J. Physiol. Paris 101: 169–178.

Vail, A.L., A. Manica and R. Bshary. 2013. Referential gestures in fish collaborative hunting. Nat. Commun. 4: 1765.

Wyatt, T.D. 2017. Animal Behavior: A Very Short Introduction. Oxford Univ. Press, Oxford, U.K.

Yuste, R. 2024. Breaking the neural code of a cnidarian: learning principles of neuroscience from the "vulgar" Hydra. Curr. Opin. Neurobiol. 86: 102869.

3

Neurons and Synapses

Elementary Units of The Nervous System

In this chapter, I would like to explain the functional units of the brain. These units, neurons and synapses are building components of the objective physicochemical structures of the brain which have been studied intensively. The principles of how they work have been elucidated to a large extent, although there are still some remaining questions and unexpected new findings are being continuously reported.

Neurons are cells which are specifically differentiated to serve information transmissions in animals and are the main components of nervous systems including the brain. Neurons are connected with other neurons, sensory cells, muscle cells and/or secretory cells at specialized structures called synapses. The primary functions of a neuron are (i) to receive information from other neurons or sensory cells at input synapses, (ii) to conduct information within a neuron toward output synapses, (iii) to transmit information to other neurons, muscle cells and secretory cells at output synapses.

There are various types of neurons showing different morphology, although all of them have two types of branching processes. Using a representative example of neurons and synapses schematically shown in Figure 1, I would like to explain the morphology and the functions of neurons further. One type of process is relatively short, and tapers and branches a number of times near the cell body, where a nucleus is localized. These types of processes are called dendrites and work as sites to receive synaptic inputs together with a cell body. Another type of process is a single axon, which is usually long and shows no tapering; in other words, the diameter does not change throughout the whole axon length. It branches at sites far from the cell body in most cases. At the ends of an axon, synapses are formed on dendrites and/or cell bodies of other neurons, etc., and these axon terminals work as output machineries. The length of an axon in some neurons such as motor neurons in the spinal cord can reach about 1m in a large person. A neuron receives synaptic inputs on dendrites and/or a cell body, and then these input signals are integrated into a cell body, more precisely at the site from which an axon stems out, that is called the axon hillock. If integrated synaptic inputs are large enough, the information is sent to axon terminals through the axon, and outputs to the next neurons at synapses.

DOI: 10.1201/9781003518822-3

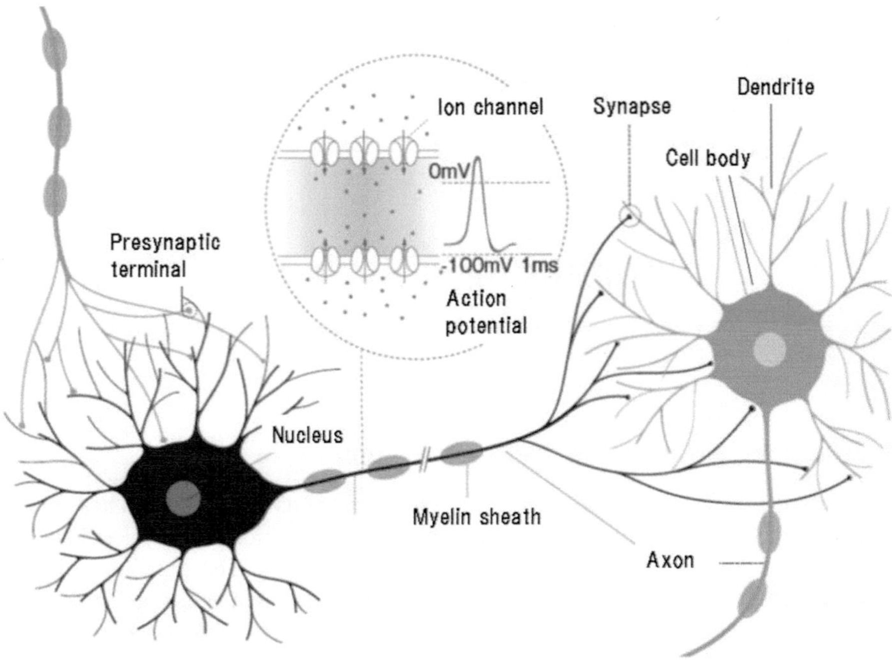

Figure 1. Scheme of neurons and synapses. A neuron receives synaptic inputs on dendrites or on a cell body. Postsynaptic membrane potential change induced by synaptic inputs are integrated in a cell body. When integrated depolarization of a cell body exceeds the threshold membrane potential, an action potential is generated through opening of sodium ion channel on an axon. The action potential moves through the axon toward presynaptic terminals, which form synapses on other neurons and secrete the neurotransmitter toward them. Myelin sheath surrounding an axon speeds up conduction of an action potential (modified from Hirano 2019).

Glial cells are another type of cells in the nervous system and outnumber neurons and support their functions. Glial cells are classified into astrocytes, oligodendrocytes, Schwann cells and microglia (Jessen 2004, Luo 2021). Astrocytes maintain appropriate concentrations of ions and neurotransmitters in the neuronal environment, and are essential regulators of formation, maintenance and the function of synapses. Oligodendrocytes and Schwann cells form myelin sheaths around axons (Figure 1), which contribute to the fast conduction of an action potential, which is explained in the next section. Microglia are phagocytes responsible for removing debris during development and in disease (Printz et al. 2019). I note that additional roles of glial cells have been reported (Jessen 2004, Allen and Lyons 2018).

Electrical Information Transmission Using Membrane Potential

Information transfer from a cell body to presynaptic terminals through an axon occurs electrically. The change of the membrane potential called an action potential (Figure 1), moves toward the axon terminals. The conduction velocity of an action potential ranges from 0.1 to 100 m/sec depending on the type and the diameter of an axon. Myelinated axons which are covered with a myelin sheath (Figure 1), show faster conduction velocity of an action potential than non-myelinated axons. Axons with larger diameters conduct action potentials faster (Luo 2021, Purves et al. 2019). The myelin sheath, primarily composed of lipid membrane formed by oligodendrocytes or Schwann cells, prevents transport of ions through the axonal membrane, and extends the length of an axonal part whose membrane potential is influenced by an action potential. The membrane potential is the electrical potential across the cell membrane, and the value inside of a cell against that of the outside is referred.

Then, how is membrane potential brought about? It is primarily caused by the uneven distribution of inorganic ions, potassium and sodium ions in particular, and by the selective permeability of the cell membrane to specific ions. The cytosol of most cells, including neurons, contains a much higher concentration (more than 10 times) of potassium ions and much less concentration (around 0.1 times) of sodium ions compared to the extracellular solution (Figure 2A). In a resting state or in an unstimulated condition, potassium ions selectively permeate the cell membrane of a neuron. As the inside concentration of potassium ions is higher and they permeate the cell membrane, potassium ions diffuse to the outside of a neuron. When some potassium ions are transferred to the outside, the inside electrical potential that is equal to the membrane potential, becomes negative as potassium ions are charged positive. This negative potential attracts potassium ions and works to keep potassium ions inside a neuron. At a certain membrane potential, which is called the equilibrium potential, the outward movement of potassium ions by diffusion and the inward movement by the electrostatic force are balanced (Figure 2A). This equilibrium potential for potassium ions is close to the resting membrane potential, which is about –60 mV in most neurons.

When the membrane potential becomes less negative, in other words when it is depolarized by synaptic inputs in most cases, the membrane potential shows a transient abrupt change to a positive potential of 10 to 40 mV. This membrane potential change is called an action potential, and it lasts about 1 milli-second (msec) (Figure 1, 2B). The action potential is brought about by the transient increase of membrane permeability to sodium ions. As the concentration of sodium ions is much higher at the outside of a neuron, sodium ions come into a neuron, resulting in the change of the membrane potential to a positive value. The reason why an action potential is transient, is that the permeability increase to sodium ions is short-lasting and the additional increase of the permeability to potassium ions follows (Figure 2B). Such membrane permeabilities to specific ions are brought about by a group of membrane proteins called ion channels, which are explained in the next section.

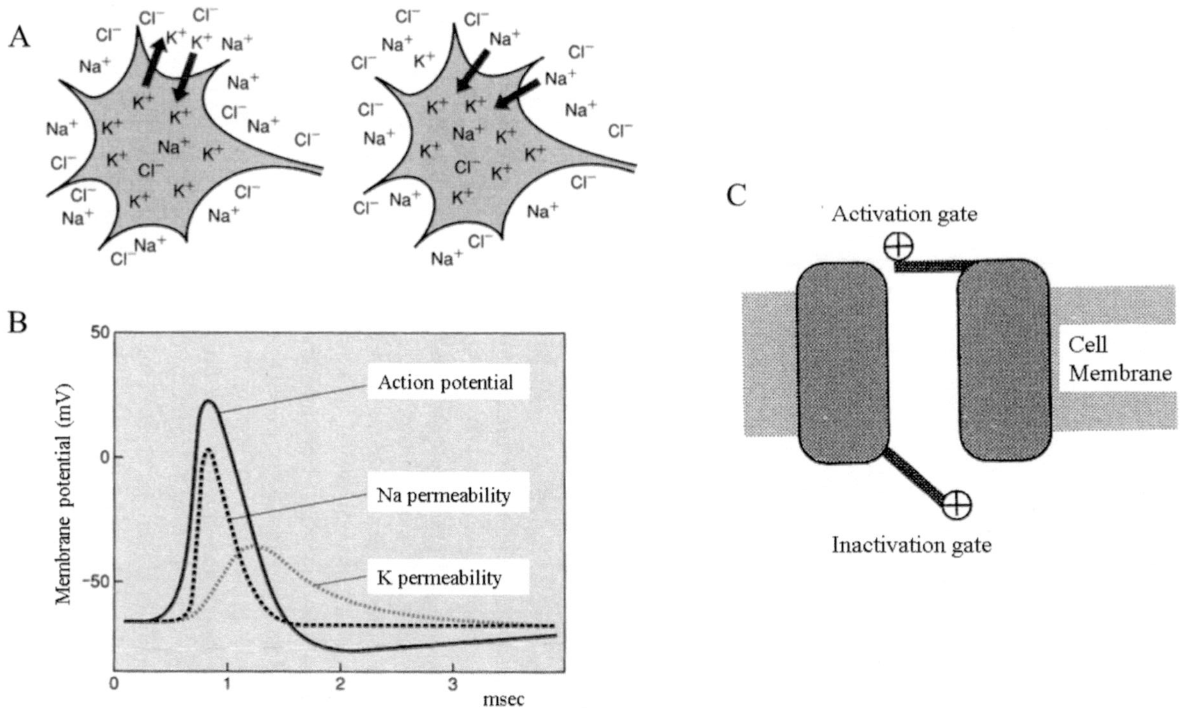

Figure 2. Resting potential and action potential. A left: Resting membrane potential of around –60 mV is caused by the uneven distribution of potassium ions (K^+) across the cell membrane and by its selective permeability to potassium ions. A right and B: Action potential is caused by a transient increase in the membrane permeability to sodium ions (Na^+), and terminates by its decrease and by the delayed increase in the permeability to potassium ions. C: A scheme of voltage-gated sodium ion channel. It has a pore, activation gates and an inactivation gate. The gate is opened or closed by electrical potential across the cell membrane through its influence on charged amino-acid residues of a channel protein. Most types of voltage-gated ion channels including sodium ion channel have more than one activation gate (modified from Hirano 2001).

Ionic Channels and Pumps

The permeability of the cell membrane at a resting state is caused by potassium ion channels which open at a resting membrane potential. I note that the major constituents of the cell membrane are lipids which form bilayers and do not allow permeation of ions (Alberts et al. 2023). When the membrane potential is depolarized, sodium ion channels activate or open, resulting in the generation of action potential. An ion channel is a membrane protein containing a pore through which ions go through, and gate structures that open or close the pore (Figure 2C) (Hille 2001). The sodium ion channel is opened or gated by the change of the membrane potential, and classified to voltage-gated ion channels. The sodium ion channel also has another gate structure called the inactivation gate that closes the pore more slowly by depolarization than the activation gate opens. The inactivation of the sodium ion channel contributes to the termination of an action potential. Slowly activating voltage-gated potassium channels, which are different from the potassium ion channels that open at a resting potential, also work to terminate an action potential (Figure 2B).

The gating properties of voltage-gated sodium and potassium ion channels determine the characteristics of an action potential such as the all-or-none law, and the possession of a threshold membrane potential and a refractory period (Hodgkin and Huxley 1952). The all-or-none law of action potential is that all action potentials in a neuron show the same amplitude and the same time course, and such an action potential is generated when the membrane potential changes to the threshold membrane potential, which is in many cases about 10 mV more positive than the resting membrane potential. A refractory period is the time after an action potential, during which another action potential cannot be generated. Thus, the activities of neurons are digital in principle: action potentials show all-or-non properties, and the strength of information is expressed not by its amplitude but by the number and the frequency of action potentials.

There are different types of voltage-gated ion channels (Hille 2001). In particular, there are many types of potassium ion channels showing different membrane-potential dependences, different opening time courses, additional dependence of gating to intracellular calcium ions, etc. There are also voltage-gated calcium ion channels that selectively permeate divalent calcium cations and chloride ion channels that permeate chloride anions. These channels also show diversities. The combination of expressed voltage-gated ion channels on a neuron determines its activity pattern. Neurons possessing certain combinations of ion channels are intrinsically active: they generate action potentials without any synaptic inputs (Huguenard and McCormick 1992, Raman and Bean 1999). Such neurons contribute to the rhythmic activities of neuronal networks. Rhythmic activities are observed in breath controlling neuronal networks, locomotion controlling systems, and their involvement in higher brain functions such as perception and recognition has also been suggested (Kiehn 2006, Wyart 2018, Benedetto 2019, Ashhad et al. 2022). Intrinsically active neurons may play critical roles in breathing, dream generation during sleep, etc.

There are other types of ion channels. Some are gated by binding extracellular or intracellular chemicals. These are classified into chemically-gated ion channels. Ion channels gated by an extracellular chemical play an important role in synaptic transmission, which is explained in the following section. There are also ion channels gated by mechanical forces, which are involved in sensory transduction such as audition and detection of somatosensory stimuli. Other types of ion channels called transient receptor potential (TRP) channels are involved in temperature sensation, pain detection, etc. There are also proton ion channels, which play a critical role in sour taste perception. Sensory transduction mechanisms using these ion channels are explained further in chapter four.

Concentration differences of ions across the cell membrane are necessary for ion channels to keep or change membrane potential effectively. As explained above, the intracellular concentration of potassium ion is more than ten times of the outside, and the intracellular concentrations of sodium ions and chloride ions are about 0.1 times of the outside. As for calcium ions, the intracellular concentration is less than 0.01% of the outside concentration of about 1 milli molarity (mM). How are such large concentration differences between the inside and outside of a neuron brought about? The most important molecular machinery to cause concentration differences is a membrane protein named sodium-potassium pump or sodium-potassium ATPase (Tian and Xie 2008). It pumps out sodium ions from the inside of a cell and pumps in potassium ions, using the chemical energy provided by breaking down adenosine tri-phosphate (ATP) to adenosine di-phosphate and phosphate. ATP is produced utilizing energy liberated by the breakdown of nutrients such as carbohydrates, sugars, etc. (Alberts et al. 2023). Sodium-potassium ATPase uses about one third of the energy in the whole brain. Thus, we use a lot of energy to maintain concentration differences of ions across the neuronal membrane. There are also calcium pumps and chloride pumps (Clapham 2007, Stauber and Jentsch 2013). In addition, there are membrane proteins called exchangers including sodium-calcium exchanger. The sodium-calcium exchanger couples the influx of sodium ions to efflux of calcium ion without directly using chemical energy of breaking down ATP (DiPolo and Beaugé 2006).

Synaptic Transmission

A pair of neurons transmit information in one direction at synapses. Upon arrival of an action potential at an axon terminal, which is also called a presynaptic terminal (Figure 1), calcium ion channels on the terminal cell membrane open and the intra-terminal concentration of calcium ions increases. The increased calcium ions bind to intracellular protein named synaptotagmin, which triggers the fusion of a small intracellular vesicle (synaptic vesicle) of about 50 nano-meter (nm) diameter to the cell membrane of a presynaptic terminal, and the intra-vesicular content is released to the extracellular space between the two neurons (Figure 3) (Luo 2021, Purves 2019). Synaptic vesicles contain a high concentration of a small chemical called neurotransmitter. The presynaptic cell membrane is closely apposed (20–30 nm distance) to the postsynaptic cell membrane, and the extracellular space between the pre- and postsynaptic membrane is called the synaptic cleft. The released neurotransmitters diffuse in the synaptic cleft and some reach the postsynaptic membrane. Chemically-gated ion channels are located on the postsynaptic membrane, and they open when they are bound by the neurotransmitter molecule.

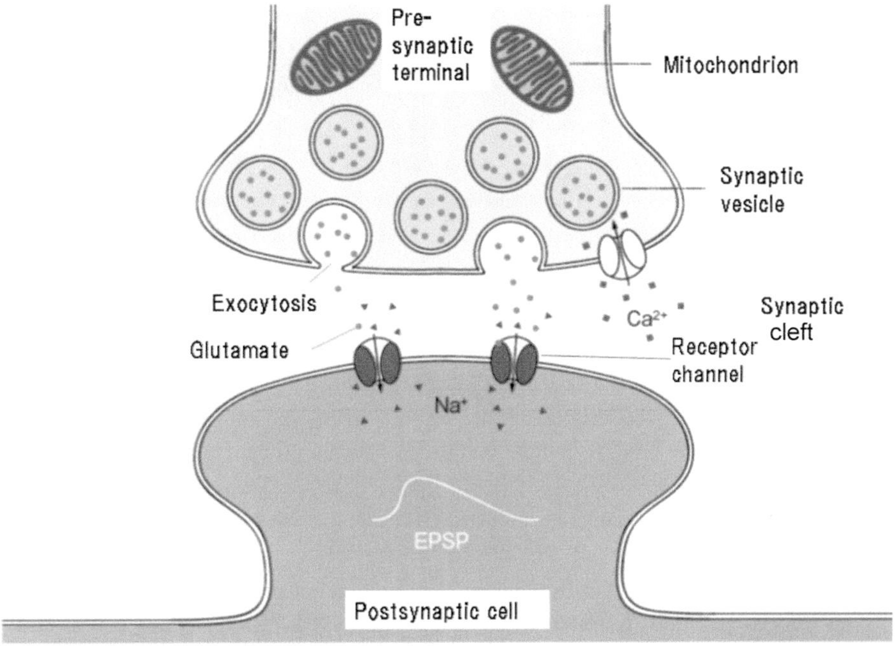

Figure 3. A scheme of excitatory synaptic transmission. Upon arrival of an action potential at a presynaptic terminal, voltage-gated calcium ion channels open and the intra-terminal concentration of calcium ions increases. Increased intra-terminal calcium ions trigger the fusion of a synaptic vesicle to the presynaptic membrane. The fusion causes the release of neurotransmitter glutamate from a synaptic vesicle to the synaptic cleft. The binding of glutamate to its postsynaptic receptor opens the pore within the receptor, through which sodium ions come into a neuron, and depolarizing EPSP is generated (modified from Hirano 2019).

Synaptic transmission takes 0.1 to 1 msec. Chemically-gated ion channels are not only ion channels but also receptor molecules. The close apposition of presynaptic and postsynaptic cell membranes is formed and maintained by synaptic adhesion molecules, which are also membrane proteins (Sűdhof 2021). They serve to gather other synaptic proteins such as those necessary for the fusion of synaptic vesicles to the presynaptic membrane, receptor ion channels on the postsynaptic membrane, etc.

In the majority of synapses in the brain, the neurotransmitter molecule is glutamate, an acidic amino-acid (Figure 4). Both sodium and potassium ions permeate glutamate-receptor ion channels. However, at a resting membrane potential mainly sodium ions come into a postsynaptic cell resulting in depolarization. The reason why mainly sodium ions move in and potassium ions do not move, is that the resting potential is near the equilibrium potential of potassium ions, at which the inflow and outflow of potassium ions are equal.

The depolarizing synaptic potential is called excitatory postsynaptic potential (EPSP), because it contributes to the generation of an action potential. The time course of EPSP is slower than that of an action potential, and lasts about several msec. EPSP caused by a single synapse in the brain or in the spinal cord are usually quite

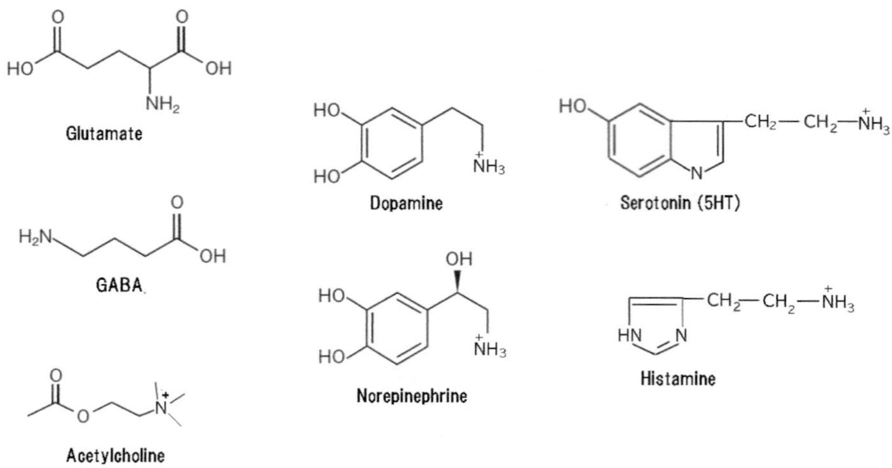

Figure 4. Chemical structures of major neurotransmitters and neuromodulators. Glutamate, GABA, acetylcholine, dopamine, norepinephrine, serotonin (5HT) and histamine are shown (Purves et al. 2019).

small (less than 1 mV), and not sufficient to trigger an action potential, because about 10 mV depolarization is required for action potential generation in most cases. Thus, multiple synaptic inputs and/or repetitive synaptic inputs that cause summation of EPSPs are necessary to trigger the generation of an action potential in a postsynaptic neuron.

Inhibitory Synapses

There are other types of synapses called inhibitory synapses. At inhibitory synapses, the neurotransmitter is either gamma-aminobutyric-acid (GABA) or amino acid glycine. In the brain GABA is prevalent (Figure 4). Not cations but chloride anions permeate through receptor ion channels of GABA or glycine. In most neurons the equilibrium potential for chloride ions is more negative than the resting potential. Therefore, the opening of these receptor channels triggered by the binding of GABA or glycine results in the increase of the negative membrane potential, which is called hyperpolarization. The hyperpolarization cancels depolarization caused by EPSPs and works to suppress action potential generation. Thus, this type of postsynaptic membrane potential change is called inhibitory postsynaptic potential (IPSP). The opening of chloride ion channels also decreases the electrical resistance of the postsynaptic membrane, in other words it forms a shunting pathway of electrical charges. This effect also works to suppress action potential generation through decreasing the amplitudes of EPSPs (Luo 2021).

It is to be noted that a neuron secretes only one kind of neurotransmitter such as glutamate or GABA in principle. As stated above, glutamate works at excitatory synapses and GABA at inhibitory synapses. Therefore, a presynaptic neuron's effect to postsynaptic neurons is either excitatory or inhibitory; in other words, neurons are classified into either excitatory neurons or inhibitory neurons in principle. However, it is to be noted that some neurons such as those secreting neuromodulator molecules have multiple neuromodulators and/or neurotransmitters, which are explained further in chapters eight and nine. In such a neuron, secretion of a neurotransmitter and that of a neuromodulator are differently regulated (Wallace and Sabatini 2023).

Transmitters and Receptors

There is another type of neurotransmitter. At the synapses between a motor neuron and a muscle cell, which is called the endplate, acetylcholine is used as the neurotransmitter (Figure 4). Cations permeate through the acetylcholine receptor ion channels that cause depolarization of the postsynaptic muscle cell; this is called endplate potential (EPP). EPP is much larger than EPSP in the brain, partly because many endplates are formed by an axon of motor neuron. Thus, EPP always triggers an action potential generation in a muscle cell, which elicits contraction through the increase in intracellular calcium ion concentration (Luo 2021). Acetylcholine is also used in some synapses in the brain. Acetylcholine is explained further in chapter nine.

I would like to note that there are significant diversities in receptor ion channels. Here, I explain glutamate receptor ion channels which are called ionotropic glutamate receptors. Ionotropic glutamate receptors are classified into N-methyl-D-aspartate (NMDA) receptors and non-NMDA receptors. Non-NMDA receptors are further classified to α-3-hydroxy-5-methyl-4-isoxazole propionic acid (AMPA) receptors and kainate receptors (Kumar and Mayer 2013). NMDA, AMPA and kainate are names of agonists that specifically bind to each type of ionotropic glutamate receptors. AMPA receptors are most prevalent and contribute to EPSP generation, whereas kainate receptors are less prevalent, but also contribute to EPSP generation at some synapses. Only monovalent cations such as sodium and potassium ions permeate most of AMPA and kainate receptor ion channels, although calcium ions permeate a few types of AMPA receptors. On the other hand, NMDA receptor channels show different characteristics such as permeability to calcium ions and pore block by a magnesium ion (Nowak et al. 1984). These characteristics play important roles in the induction of synaptic plasticity which is explained in a later section. There are several subtypes in each AMPA, kainate and NMDA receptors.

All ionotropic glutamate receptors are composed of four subunits, in other words four similar poly-peptides form a receptor ion channel with a pore at the center. Different combinations of subunits give rise to somewhat different functional properties (Kumar and Mayer 2013). Each acetylcholine and GABA receptor is composed of five subunits, and they also show significant diversities (Hogg et al. 2003, Sallard et al. 2021).

There are other types of neurotransmitters and/or neuromodulators such as amines and peptides (Figure 4). Most of them bind to another type of receptors called metabotropic receptors, although some of them also bind to ionotropic receptors (Luo 2021). Metabotropic receptors are coupled with heterotrimeric guanine-nucleotide-binding regulatory proteins (G-proteins), and exert effects through intracellular molecular signaling cascades, which are described in the next section. There are also metabotropic glutamate receptors (Niswender and Conn 2010), GABA receptors and acetylcholine receptors (Bowery et al. 2002, Dhein et al. 2001). Some amine transmitters such as norepinephrine, dopamine, histamine and serotonin are often secreted toward wider extracellular spaces and influence functions of a relatively large number of neurons simultaneously, which are called volume transmission (Özçete et al. 2024). Such amines are better to be classified to neuromodulators rather than typical neurotransmitters. Neuromodulators are explained in chapter nine.

Slow Modulation of Synaptic Functions and Neuronal Activities

The activation of metabotropic receptors influences synaptic activity and/or neuronal excitability in a slower time course and for a longer time than ionotropic receptors, because they change activities of enzymes such as phospholipase C, adenylyl cyclase, etc. through G proteins, and these changes last longer. The activities of such enzymes influence the concentration of small intracellular messenger molecules such as inositoltriphosphate (IP3) and cyclic adenosine monophosphate (cAMP), and also membrane-bound diacylglycerol (DAG). In turn, these small messenger molecules, which are called second messengers, alter activities of ion channels directly or indirectly through activation of other enzymes such as protein kinases (Alberts et al. 2023). More detailed explanation of intracellular chemical signaling pathways are explained in the next section.

Metabotropic acetylcholine receptors are known as muscarinic receptors, which are found not only in neurons but also in heart cells. Parasympathetic vagal nerves release acetylcholine to heart cells, and affect the heart rate, etc. through the activation of muscarinic receptors (Dhein et al. 2001). Metabotropic GABA receptors are known as $GABA_B$ receptors (Bowery et al. 2002), and found in some presynaptic terminals and inhibit neurotransmitter release through suppression of voltage-gated calcium ion channels. In some postsynaptic neurons, the $GABA_B$ receptor activation increases the membrane permeability to potassium ions and causes hyperpolarization with a slow time course. Metabotropic glutamate receptors also affect neuronal and synaptic activities for longer times than ionotropic receptors in various ways (Niswender and Conn 2010.). Among aminergic receptors, dopamine receptors mainly work through up- or down-regulation of adenylyl cyclase activity (Missale et al. 1998). β-adrenergic receptors are found not only in neurons but also in the heart, and work through the activation of adenylyl cyclase (Wallukat 2002), whereas α1-adrenergic receptors work mainly through regulation of phospholipase C activity (Alberts et al. 2023). Adrenergic receptors are explained further in chapter nine.

Metabotropic receptors are also called G-protein coupled receptors (GPCRs). There are many different types of GPCRs. Among them there are about 1000 types of GPCRs bound by various odorants on the cell membrane of olfactory receptor neurons in the nose. GPCRs also play crucial roles in light detection, and bitter, sweet and umami tastes sensations. Thus, GPCRs are crucial in sensory systems, which are explained in chapter four. Receptors of neuromodulator molecules are mostly GPCRs and play critical roles in emotions, attention, etc. which are explained further in chapters eight and nine.

Intracellular Signaling Cascade

G proteins, coupled with GPCRs, are heterotrimeric membrane protein composed of α, β and γ subunits (Syrovatkina et al. 2016, Alberts et al. 2023). Upon the activation of a GPCR, α subunit is released from βγ subunits in the cell membrane, and activates or suppresses membrane bound enzymes such as adenylyl cyclase or phospholipase C. G proteins activating adenylyl cyclase are classified to Gs type, and those inhibiting are Gi type. G proteins activating phospholipase C are classified to Gq type.

Adenylyl cyclase is an enzyme that breaks down ATP to cAMP and diphosphate, and cAMP acts as a second messenger molecule in the cytosol. cAMP binds to and activates A kinase, which phosphorylates some serine or threonine residues of various proteins including ion channels, and regulates their functions (Alberts et al. 2023). An example of signaling cascade including Gs, adenylyl cyclase, cAMP, A kinase and potassium ion channel is explained in the next section. Phospholipase C is an enzyme that hydrolyses phosphatidylinositol 4,5-bisphosphate to IP3 and DAG. IP3 is released to the cytosol and binds to IP3 receptor on the membrane of endoplasmic reticulum inside a cell, which includes a higher concentration of calcium ions than the cytosol. IP3 receptor is actually a calcium ion channel opened by IP3 binding (Mikoshiba and Hattori 2000). Thus, the activation of Gq protein causes an increase in the intracellular calcium ion concentration. Calcium ions together with DAG activate protein kinase C, which phosphorylates some serine or threonine residues of various proteins.

Intracellular calcium ion is also an important second messenger (Clapham 2007, Alberts et al. 2023). There are a number of proteins activated by the binding of calcium ions such as synaptotagmin, which is involved in neurotransmitter release, calmodulin, etc. Calmodulin bound by calcium ions can activate another type of serine and threonine protein kinase called calcium and calmodulin dependent kinase (CaMK). The intracellular increase in calcium ion concentration plays important roles in various cellular functions such as the secretion of neurotransmitters or hormones, muscle contraction, etc. I note that there are other intracellular signaling pathways using other types of molecules such as tyrosine kinases, phosphatases, monomeric G proteins, etc., and some pathways regulate transcription of messenger ribonucleic acid (mRNA) and synthesis of proteins (Alberts et al. 2023). Intracellular molecular signaling cascades are complicated with activation, inhibition, modulation, divergence, convergence and loop pathways, and they sometimes show non-linear outputs with thresholds in input signal intensities and all-or-none like responses, etc. (Bhalla and Iyengar 1999, Kitagawa et al. 2009).

Synaptic Plasticity

Synaptic plasticity is neuronal activity-dependent modulation of the efficacy of synaptic transmission. Synaptic plasticity has been considered as a primary cellular basis of learning and memory, because changes in synaptic transmission efficacy caused by neuronal activities accompanied with some experiences can alter following behaviors (Okano et al. 2000, Luo 2021, Purves et al. 2019).

A molluscan species, called sea hare *Aplysia,* possesses a simple nervous system composed of large identifiable neurons. Behavioral changes in *Aplysia,* caused by experiences have been directly explained by changes of efficacy of synaptic transmission. An *Aplysia* shows a simple defensive reflex called gill withdrawal reflex, which is shrinkage of gills caused by the application of a mechanical stimulation around the gills, such as a siphon (Figure 5A) (Carew et al. 1971). Gill withdrawal reflex is mediated only by two types of neurons, mechano-sensory neurons responding to mechanical stimulation to the siphon and motor neurons controlling muscles of the gills. Repetition of the mechanical stimulation gradually decrease the amplitude of gill retraction accompanied with the decrease in amplitude of EPSP in the motor neuron caused by the sensory neuron. This behavioral change is called habituation, and the occurrence mechanism is the decrease of EPSP amplitude caused by suppressed neurotransmitter release (Castellucci and Kandel 1974), which is ascribed to the decreased calcium ion influx to the presynaptic terminal. The significance of habituation has been suggested to be effort saving by suppressing unnecessary behavioral action to harmless stimuli.

As for the gill withdraw reflex of *Aplysia*, another phenomenon called sensitization has also been reported (Figure 5A) (Castellucci and Kandel 1976). When another relatively strong mechanical stimulation is applied to Aplysia on the region distant from the gills such as a tail, the amplitude of gill-withdrawal reflex triggered by the siphon stimulation is increased for a while. This phenomenon is called sensitization, and can be interpreted to say that the *Aplysia* becomes more cautious after receiving stimulus to the tail and comes to respond to possibly harmful stimulations more strongly. The stimulation to the tail is detected by tail sensory neurons, which send the information to an interneuron forming synapses on the presynaptic terminal of a siphon sensory neuron (Figure 5A). The interneuron secretes serotonin, a type of amine neurotransmitter (Figure 4), that increases the transmitter release from the presynaptic terminal of the siphon sensory neuron to the motor neuron.

The molecular control mechanism that shows how serotonin increases transmitter release has been elucidated (Figure 5B) (Kandel and Schwartz 1982). It is an example of intracellular signaling cascade including GPCR, G protein, adenylyl cyclase, cAMP, A kinase, potassium ion channel and calcium ion. Serotonin binds to the metabotropic serotonin receptor coupled to Gs protein, which activates adenylyl cyclase. Adenylyl cyclase produces cAMP, which binds and activates A kinase. One substrate of A kinase is a type of potassium ion channel, whose activity is suppressed by the phosphorylation. The decreased activity of the potassium ion channel works to increase the depolarization of the presynaptic terminal of the sensory neuron, and increases the calcium influx through voltage-gated calcium ion channel. Then, the increased intracellular calcium ion concentration enhances the transmitter release onto the motor neuron, resulting in augmented gill withdrawal reflex. Thus, in *Aplysia* the cellular and molecular mechanisms of simple forms of learning and memory have been elucidated, and synaptic plasticity has been identified as a key mechanism of learning.

A psychologist Donald Hebb suggested that synaptic plasticity is critical for learning and memory, and proposed what is known as Hebb's rule (Hebb 1949).

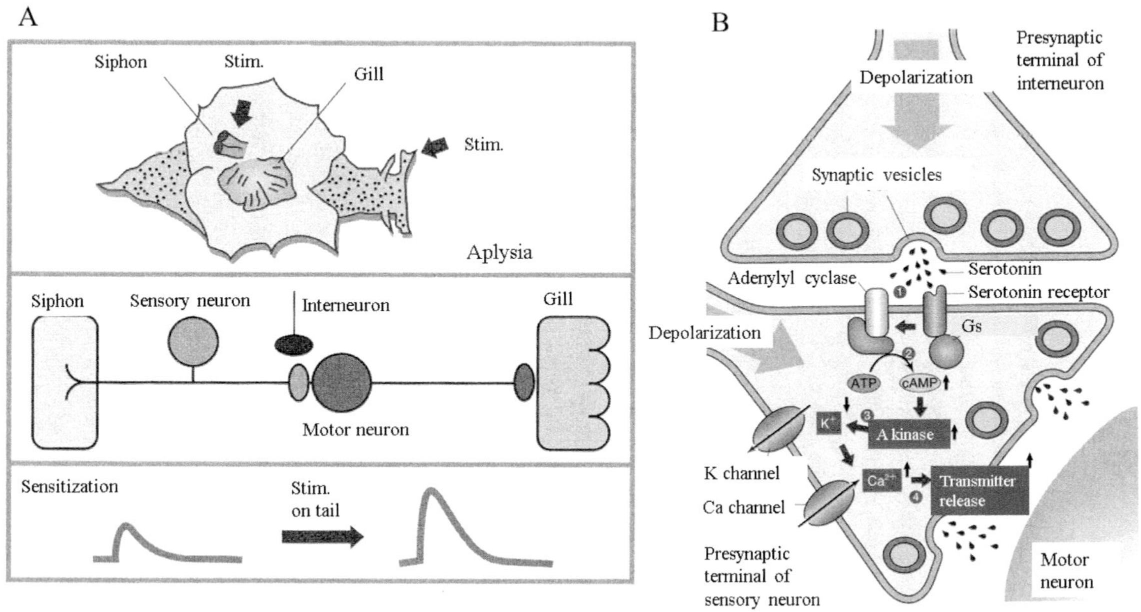

Figure 5. Schematic presentation of gill withdrawal reflex in *Aplysia* and its sensitization. A: *Aplysia* (top), and the neuronal pathway of gill withdrawal reflex consisting of sensory neurons responding to mechanical stimulation to the siphon and motor neurons controlling muscles of the gills (middle). Sensitization increases the amplitude of EPSP in a motor neuron and enhances the reflex (bottom). B: Molecular induction mechanism of presynaptic facilitation of transmitter release at a synapse between the sensory neuron and the motor neuron, which underlies sensitization of gill withdrawal reflex in *Aplysia*. Serotonin released from an interneuron binds to the metabotropic receptor coupled to Gs protein (1), which activates adenylyl cyclase in the presynaptic terminal of the sensory neuron. Adenylyl cyclase produces cAMP (2), which binds and activates A kinase. A kinase phosphorylates a type of potassium ion channel and suppresses its activity (3), resulting in enhanced depolarization of the presynaptic terminal of the sensory neuron, which in turn increases the calcium influx through voltage-gated calcium ion channel. The increased intracellular calcium ion concentration enhances the transmitter release onto the motor neuron (4), and increases the amplitude of EPSP, resulting in enhancement of the gill withdrawal reflex (modified from Hirano 2001, Carew et al. 1971, Kandel and Schwartz 1982).

It said, "When an axon of cell A is near enough to excite a cell B and repeatedly or persistently takes part in firing it, some growth process or metabolic change takes place in one or both cells such that A's efficiency, as one of the cells firing B, is increased." Synaptic plasticity reminiscent of Hebb's rule has been reported at various synapses.

Actually, any synapses show some forms of plasticity, and synaptic plasticity can be classified as potentiation and depression. Among them the long-lasting ones seem to be more important for memory. Long-term potentiation (LTP), that is sustained enhancement of efficacy of synaptic transmission, has been reported to occur in various brain regions such as the hippocampus, the cerebral cortex, etc. after a high frequency repetitive activation of the synapse (Figure 6A) (Nicoll 2017). On the other hand, a low frequency activation of synapses induces long-term depression (LTD) at some synapses (Bear and Abraham 1996).

The molecular mechanisms of induction of LTP and LTD have been intensively studied (Nicoll 2017, Bear and Abraham 1996). The critical contribution of NMDA-type glutamate receptor ion channel has been clarified. Upon repetitive activation of a synapse, an EPSP is followed by the next EPSP before the former comes back to the basal level of membrane potential, resulting in the accumulation of depolarization which is called temporal summation (Figure 6A). The relatively large depolarization caused by temporal summation of EPSP repels the magnesium ion block of the NMDA receptor channel, causing the calcium ion influx (Figure 6B). Increased intracellular calcium ions bind to calmodulin and activate CaMKII, which phosphorylates AMPA-type glutamate receptors, etc., resulting in an increase in the number of AMPA receptors on the postsynaptic membrane, which increases the amplitude of EPSP. The increase in the size of postsynaptic structure called a spine during LTP was also reported (Matsuzaki et al. 2004). The causal link between the hippocampal LTP and learning has been studied using mutant mice defective in LTP, etc., which suggest that they are related (Lynch 2004, Nicoll 2017).

As mentioned above, LTD is induced by low frequency activation of some synapses. Here, a relatively low-level increase of intracellular calcium ion concentration caused by low-level activation of NMDA receptor, activates a type of phosphatase calcineurin more than CaMKII. Phosphatase is an enzyme that dephosphorylates substrates and counteracts the function of kinases, and works to decrease the amplitude of EPSP (Alberts et al. 2023).

Bidirectional synaptic plasticity depending on the timings of action potentials, in both pre- and postsynaptic neurons, which is called spike timing dependent plasticity (STDP) has also been reported (Debanne and Inglebert 2023). In STDP, potentiation is induced when a presynaptic action potential is followed by the generation of a postsynaptic action potential in a few msec repetitively, and depression is induced when a postsynaptic action potential precedes a presynaptic action potential. Thus, STDP is characterized by an asymmetric plasticity rule that could be interpreted to assume that the direction of the plasticity is determined by whether the presynaptic activity can be the cause of postsynaptic activity or not, which is reminiscent of Hebb's rule. The effectiveness of synaptic activity determines the direction of plastic change in STDP.

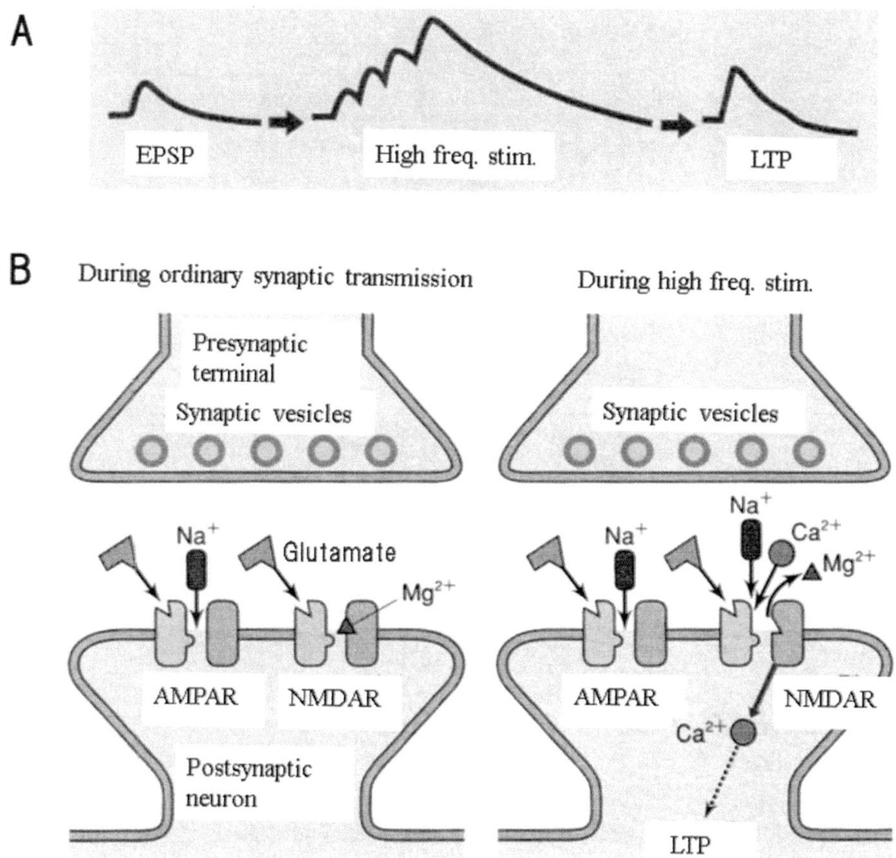

Figure 6. Induction of LTP by repetitive activation of the synaptic activity. A: The temporal summation of EPSP causes large depolarization in a postsynaptic neuron, which induces LTP. B: The induction mechanism of hippocampal LTP. In a base line glutamatergic synaptic transmission, the amplitude of EPSP caused by the opening of AMPA-type glutamate receptor is relatively small, and the NMDA-type glutamate receptor ion channel pore is blocked by magnesium ion. On the other hand, during high frequency activation of the synapse, large depolarization caused by temporal summation of EPSP repels magnesium ion from the pore of NMDA receptor channel, and calcium ions get into the postsynaptic neuron which induces LTP (modified from Hirano 2001).

A different type of LTD has been reported in the cerebellum, and it has been suggested to contribute to motor learning (Ito 2001, Hirano 2013). A cerebellar Purkinje neuron is a type of inhibitory neuron sending a sole output from the cerebellar cortex. A Purkinje neuron receives two very different types of glutamatergic synaptic inputs. One is a very strong single climbing fiber input from an inferior olivary neuron, and the other is numerous weak inputs from parallel fibers, which are axons of granule neurons (Figure 7). The transmission efficacy of the latter undergoes LTD, when the activation of these two types of synapses is coupled. This cerebellar LTD has been considered to work so that inadequate information transmitted through a group of parallel fibers are suppressed by an error signal conveyed by a climbing

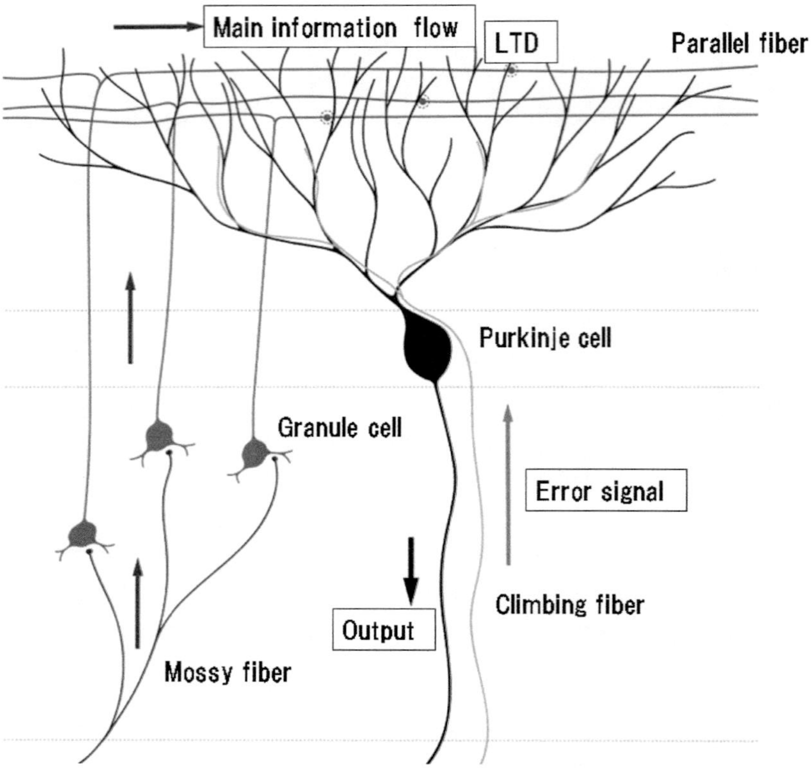

Figure 7. Cerebellar cortical neuronal circuits. A Purkinje cell is a primary neuron in the cerebellar cortex sending a sole output from the cortex. It receives a large number of excitatory glutamatergic synaptic inputs from granule neurons through parallel fibers, and a very strong synaptic input from only one inferior olivary neuron through a climbing fiber, which conveys an instructive error signal. Conjunctive activation of parallel fibers and a climbing fiber causes LTD at the parallel fibers to Purkinje neuron synapses (modified from Hirano 2019).

fiber. The molecular induction mechanism of cerebellar LTD has also been studied. A large increase in the intracellular concentration of calcium ions brought about by a climbing fiber activation together with DAG produced by activation of parallel fibers through metabotropic glutamate receptor 1 (mGluR1), Gq protein and phospholipase C, activate protein kinase C, resulting in the decrease in number of AMPA-type glutamate receptors on the postsynaptic membrane. Additional explanation of the cerebellar LTD is provided in chapter six.

Induction conditions, mechanisms and expression of synaptic plasticity are diverse, and not only excitatory synapses but also inhibitory synapses show synaptic plasticity (Hirano and Kawaguchi 2014, Hennequin et al. 2017). Another plasticity mechanism known as homeostatic synaptic plasticity has also been reported (Turrigiano 2012, Chen et al. 2022). It works to stabilize neuronal activities by scaling synaptic inputs and controlling the balance of excitation and inhibition. Regulation and interaction of different types of synaptic plasticity through intracellular molecular signaling cascades are complicated (Kawaguchi and Hirano 2013).

The Control of Transcription and Long-Term Modification of Neuronal Circuits

When stimulation causing sensitization in *Aplysia* are repeated for several days, sensitization lasting for weeks can be induced. In this case, the transcription of mRNA as well as the morphology of the synapses change. Here, the enhancement of transcription is induced cAMP-dependently (Kaang et al. 1993). Longer lasting synaptic plasticity involving alteration of transcription has also been reported at other synapses (Alberini 2009, Ma et al. 2023). The postsynaptic increase in calcium ion concentration induces new gene transcription (Cohen and Greenberg 2008).

The formation of new synapses and elimination of synapses have been reported during development and after maturation (McAllister 2007, Kano and Hashimoto 2009, Choucry et al. 2024). The numbers of most types of neurons do not increase but decrease with aging. The loss of some neurons seems to be supplemented by the increase in the number of synapses which single neurons form. In the hippocampal dentate gyrus region, neurogenesis occurs even in adults, and new-born neurons are integrated into neuronal circuits (Cope and Gould 2019).

Summary and Conclusion

Neurons are building units of the brain, and transmit signals to other neurons at synapses. The primary functions of a neuron are (i) to receive information from other neurons or sensory cells at input synapses and generate postsynaptic potential, (ii) to conduct information using an action potential through an axon toward output synapses, (iii) to transmit information to other neurons or muscle cells by releasing neurotransmitter from presynaptic terminals. Both an action potential and postsynaptic potentials are changes of the membrane potential. Thus, information transfer by neurons is electrical.

The membrane potential is determined by uneven distribution of ions such as potassium, sodium, calcium and chloride ions across the cell membrane, and by selective permeability of the cell membrane caused by opening of ion channels. At a resting state the neuronal membrane potential is about –60 mV which is close to the equilibrium potential of potassium ions. When the membrane potential is depolarized to the threshold potential, the membrane potential shows an abrupt change to a positive value transiently, by the increased sodium ion permeability. This phenomenon is called an action potential, which propagates through an axon toward presynaptic terminals. In the synaptic transmission, a type of transmitter molecules is secreted from the presynaptic terminal, and the postsynaptic potential change is caused by the opening of transmitter-receptor ion channels. The postsynaptic potential change is either depolarizing (EPSP) or hyperpolarizing (IPSP), which is determined by the ion selectivity of the receptor channel's pore. In the brain, EPSP is mainly caused by the glutamate receptor ion channels permeated by cations, whereas IPSP is mainly caused by GABA receptor ion channels permeated by chloride anions. Any single neuron secretes only one type of neurotransmitter in principle. Thus, neurons are classified as either excitatory or inhibitory neurons.

There are also neurotransmitter receptors that couple not with ion channels but with heterotrimeric G proteins called metabotropic receptors or GPCRs. The activation of a metabotropic receptor causes slow modulation of neuronal and/or synaptic activities through intracellular molecular signaling cascades. The synaptic activity-dependent alteration of the efficacy of synaptic transmission is called synaptic plasticity, which has been considered as a primary cellular basis of learning and memory. Long-term potentiation (LTP) and long-term depression (LTD) has been reported in various synapses, and their induction mechanisms and roles have been intensively studied. Other types of synaptic plasticity such as spike timing dependent plasticity, homeostatic synaptic plasticity and plasticity at inhibitory synapses have also been reported.

Ion channels, receptor molecules, activity patterns of neurons, and characteristics of synaptic plasticity show significant diversities and complexity, and the numbers of synapses and neurons change significantly during life. Thus, the building blocks of the brain are complicated and flexible.

References

Alberini, C. 2009. Transcription factors in long-term memory and synaptic plasticity. Physiol. Rev. 89: 121–145.

Alberts, B., K. Hopkin, A. Johnson, D. Morgan, K. Roberts, P. Walter et al. 2023. Essential Cell Biology. 6th Ed. Norton, New York, NY.

Allen, N.J. and D.A. Lyons. 2018. Glia as architects of central nervous system formation and function. Sci. 362: 181–185.

Ashhad, S., K. Kam, C.A. Del Negro and J.L. Feldman. 2022. Breathing rhythm and pattern and their influence on emotion. Annu. Rev. Neurosci. 45: 223–247.

Benedetto, A., M.C. Morrone and A. Tomassini. 2019. The common rhythm of action and perception. J. Cogn. Neurosci. 32: 187–200.

Bear, M.F. and W.C. Abraham. 1996. Long-term depression in hippocampus. Annu. Rev. Neurosci. 19: 437–462.

Bhalla, U.S. and R. Iyengar. 1999. Emergent properties of networks of biological signaling pathways. Sci. 283: 381–387.

Bowery, N.G., B. Bettler, W. Froestl, J.P. Gallagher, F. Marshall, M. Raiteri et al. 2002. International Union of Pharmacology. XXXIII. Mammalian γ-aminobutyric acid B receptors: structure and function. Pharmacol. Rev. 54: 247–264.

Carew, T.J., V.F. Castellucci and E.R. Kandel. 1971. An analysis of dishabituation and sensitization of the gill-withdrawal reflex in *Aplysia*. Int. J. Neurosci. 2: 79–98.

Castellucci, V.F. and E.R. Kandel. 1974. A quantal analysis of the synaptic depression underlying habituation of the gill-withdrawal reflex in *Aplysia*. Proc. Nat. Acad. Sci. USA 71: 5004–5008.

Castellucci, V.F. and E.R. Kandel. 1976. Presynaptic facilitation as a mechanism for behavioral sensitization in *Aplysia*. Sci. 194: 1176–1178.

Chen, L., X. Li, M. Tjia and S. Thapliyal. 2022. Homeostatic plasticity and excitation-inhibition balance: the good, the bad, and the ugly. Curr. Opin. Neurobiol. 75: 102553.

Choucry, A., M. Nomoto and K. Inokuchi. 2024. Engram mechanisms of memory linking and identity. Nat. Rev. Neurosci. 25: 375–392.

Clapham, D.E. 2007. Calcium signaling. Cell. 131: 1047–1058.

Cohen, S. and M.E Greenberg. 2008. Communication between the synapse and the nucleus in neuronal development, plasticity, and disease. Annu. Rev. Cell Dev. Biol. 24: 183–209.

Cope, E.C. and E. Gould. 2019. Adult neurogenesis, glia, and the extracellular matrix. Cell Stem Cell 24: 690–705.
Debanne, D. and Y. Inglebert. 2023. Spike timing-dependent plasticity and memory. Curr. Opin. Neurobiol. 80: 102707.
Dhein, S., C.J. van Koppen and O.E. Brodde. 2001. Muscarinic receptors in the mammalian heart. Pharmacol. Res. 44: 161–182.
DiPolo, R and L. Beaugé. 2006. Sodium/calcium exchanger: influence of metabolic regulation on ion carrier interactions. Physiol. Rev. 86: 155–203.
Hebb, D.O. 1949. The Organization of Behavior: A Neuropsychological Theory. Wiley & Sons, New York, NY.
Hennequin, G., E.J. Agnes and T.P. Vogels. 2017. Inhibitory plasticity: balance, control, and codependence. Annu. Rev. Neurosci. 40: 557–579.
Hille, B. 2001. Ion Channels of Excitable Membranes. 3rd. Ed. Sinauer, Sunderland, MA.
Hirano, T. 2001. Identity of Brain and Mind (Japanese). Tokyo Kagakudoujin Press, Tokyo.
Hirano, T. 2013. Long-term depression and other synaptic plasticity in the cerebellum. Proc. Japan Acad. Series B 89: 183–195.
Hirano, T. 2019. What is the Brain for? Behavioral Selection and Neuroscience in the Era of AI (Japanese) Kyoto University Press, Kyoto, Japan.
Hirano, T. and S. Kawaguchi. 2014. Regulation and functional roles of rebound potentiation at cerebellar stellate cell - Purkinje cell synapse. Front. Cell. Neurosci. 8: 42.
Hodgkin, A.L. and A.F. Huxley. 1952. A quantitative description of membrane current and its application to conduction and excitation in nerve. J. Physiol. 117: 500–544.
Hogg, R.C., M. Raggenbass and D. Bertrand. 2003. Nicotinic acetylcholine receptors: from structure to brain function. Rev. Physiol. Biochem. Pharmacol. 147: 1–46.
Huguenard, J.R. and D.A. McCormick. 1992. Simulation of the currents involved in rhythmic oscillations in thalamic relay neurons. J. Neurophysiol. 68: 1373–1383.
Ito, M. 2001. Cerebellar long-term depression: characterization, signal transduction, and functional roles. Physiol. Rev. 81: 1143–1195.
Jessen, K.R. 2004. Glial cells. Int. J. Biochem. Cell Biol. 36: 861–1867.
Kaang, B.K., E.R. Kandel and S.G. Grant. 1993. Activation of cAMP-responsive genes by stimuli that produce long-term facilitation in Aplysia sensory neurons. Neuron 1: 427–435.
Kandel, E.R. and J.H. Schwartz. 1982. Molecular biology of learning: modulation of transmitter release. Sci. 218: 433–443.
Kano, M. and K. Hashimoto. 2009. Synapse elimination in the central nervous system. Curr. Opin. Neurobiol. 19: 154–161.
Kawaguchi, S. and T. Hirano. 2013. Gating of long-term depression by CaMKII through enhanced cGMP signaling in cerebellar Purkinje cells. J. Physiol. 591: 1707–1730.
Kiehn, O. 2006. Locomotor circuits in the mammalian spinal cord. Annu. Rev. Neurosci. 29: 279–306.
Kitagawa, Y., T. Hirano and S. Kawaguchi. 2009. Prediction and validation of a mechanism to control the threshold for inhibitory synaptic plasticity. Mol. Systems. Biol. 5: 280.
Kumar, J. and M.L. Mayer. 2013. Functional insights from glutamate receptor ion channel structures. Annu. Rev. Physiol. 75: 313–737.
Luo, L. 2021. Principles of Neurobiology. 2nd Ed. CRC Press, Boca Raton, FL.
Lynch, M.A. 2004. Long-term potentiation and memory. Physiol. Rev. 84: 87–136.
Ma, H., H.G. Khaled, X. Wang, N.J. Mandelberg, S.M. Cohen, X. He et al. 2023. Excitation–transcription coupling, neuronal gene expression and synaptic plasticity. Nat. Rev. Neurosci. 24: 672–692.
Matsuzaki, M., N. Honkura, G.C.R. Ellis-Davies and H. Kasai. 2004. Structural basis of long-term potentiation in single dendritic spines. Nat. 429: 761–766.
McAllister, A.K. 2007. Dynamic aspects of CNS synapse formation. Annu. Rev. Neurosci. 30: 425–450.
Mikoshiba, K. and M. Hattori. 2000. IP3 receptor-operated calcium entry. Sci. STKE 51: pe1.
Missale, C., S.R. Nash, S.W. Robinson, M. Jaber and M.G. Caron. 1998. Dopamine receptors: from structure to function. Physiol. Rev. 78: 189–225.
Nicoll, R.A. 2017. A brief history of long-term potentiation. Neuron 93: 281–290.
Niswender, C.M. and P.J. Conn. 2010. Metabotropic glutamate receptors: physiology, pharmacology, and disease. Annu. Rev. Pharmacol. Toxicol. 50: 295–322.

Nowak, L., P. Bregestovski, P. Ascher, A. Herbet and A. Prochiantz. 1984. Magnesium gates glutamate-activated channels in mouse central neurons. Nat. 307: 462–465.

Okano, H., T. Hirano and E. Balaban. 2000. Learning and memory. Proc. Natl. Acad. Sci. USA 97: 12403–12404.

Özçete, Ö.D., A. Banerjee and P.S. Kaeser. 2024. Mechanisms of neuromodulatory volume transmission. Mol. Psychiatry. doi: 10.1038/s41380-024-02608-3.

Prinz, M., S. Jung and J. Priller. 2019. Microglia biology: one century of evolving concepts. Cell 179: 292–311.

Purves, D., G.J. Augustine, D. Fitzpatrick, W.C. Hall, A.S. LaMantia, R.D. Mooney et al. 2019. Neuroscience. 6th Ed. Sinauer, Sunderland, MA.

Raman, I.M. and B.P. Bean. 1999. Ionic currents underlying spontaneous action potentials in isolated cerebellar Purkinje neurons. J. Neurosci. 19: 1663–1674.

Sallard, E., D. Letourneur and P. Legendre. 2021. Electrophysiology of ionotropic GABA receptors. Cell Mol. Life Sci. 78: 5341–5370.

Stauber, T. and T.J. Jentsch. 2013. Chloride in vesicular trafficking and function. Annu. Rev. Physiol. 75: 453–477.

Sűdhof, T.C. 2021. The cell biology of synapse formation. J. Cell Biol. 220: e202103052.

Syrovatkina, V., O. Kamela, A.R. Dey and X. Huang. 2016. Regulation, signaling, and physiological functions of G-Proteins. J. Mol. Biol. 428: 3850–3868.

Tian, J. and Z. Xie. 2008. The Na-K-ATPase and calcium-signaling microdomains. Physiol. 23: 205–211.

Turrigiano, G. 2012. Homeostatic synaptic plasticity: local and global mechanisms for stabilizing neuronal function. Cold Spring Harb. Perspect. Biol. 4: a005736.

Wallace, M.L. and B.L. Sabatini. 2023. Synaptic and circuit functions of multitransmitter neurons in the mammalian brain. Neuron 111: 2969–2983.

Wallukat, G. 2002. The beta-adrenergic receptors. Herz 27: 683–690.

Wyart, C. 2018. Taking a big step towards understanding locomotion. Trends Neurosci. 41: 869–870.

4

The Functions of Neuronal Circuits

Sensory Processing and Motor Control

Peripheral and Central Nervous Systems

Nervous systems are divided into the central nervous system (CNS) and the peripheral nervous system (PNS). The brain and the spinal cord are CNS, where a large number of neuronal cell bodies are localized, and both sensory information processing and motor command generation are carried out. In addition, the brain performs higher information processing such as perception, recognition, planning, memory, emotions, etc. Functions of the brain seem to underly self, mind and consciousness. During ontology the brain and the spinal cord are formed from the neuronal tube.

PNS is composed of axon bundles and ganglia such as the dorsal root ganglia and the sympathetic ganglia. Ganglia contain neuronal cell bodies sending axons to peripheries of the body. The dorsal root ganglia contain sensory neurons connecting peripheral mechano-sensory cells, etc. and neurons in the spinal cord. Axon bundles connecting the periphery and the spinal cord or the medulla oblongata are called nerves such as the sciatic nerve, the vagal nerve, the optic nerve, etc. Some nerves such as the sciatic nerve contain not only sensory axons but also axons of motor neurons, whose cell bodies are located in the spinal cord or the medulla oblongata. Developmentally, ganglia derive from the neural crests.

The Development of The Brain and The Spinal Cord

CNS derives from the neural tube, which is formed through invagination of the ectoderm above the notochord (Luo 2021). During ontogeny some portions of the neural tube swell, and form the telencephalon, the diencephalon, the mesencephalon, the rhombencephalon, the myelencephalon and the spinal cord (Figure 1A). The telencephalon gives rise to the cerebral cortex, the basal ganglia, the hippocampus, the amygdala, the olfactory bulb, etc. The diencephalon gives rise to the thalamus and the hypothalamus. In the mesencephalon or the midbrain, the superior colliculus, the inferior colliculus, the substantia nigra, etc. are formed. The cerebellum and the pons are formed from the rhombencephalon, and the myelencephalon becomes the

DOI: 10.1201/9781003518822-4

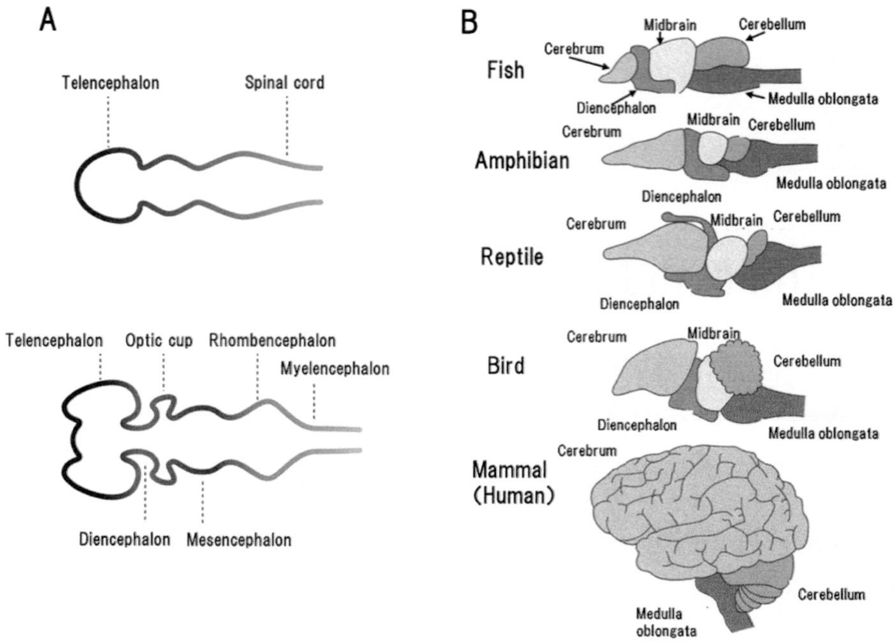

Figure 1. Development of CNS and comparison of vertebrates' CNSs. A, Formation of CNS from the neural tube. Development proceeds from the upper figure to the lower figure. In the lower figure, from the anterior (left side) the telencephalon, the diencephalon, the mesencephalon, the rhombencephalon and the myelencephalon are located. The optic nerves and the optic cups are formed from extensions of the diencephalon, and the retinae develop from the optic cups. B, Schematic presentation of vertebrates' CNSs. In a human the extensively developed cerebral cortex overlies the structures derived from the diencephalon and the mesencephalon (modified from Hirano 2021).

medulla oblongata. The mesencephalon, the pons and the medulla oblongata together are called the brainstem.

In most of vertebrates such as fishes, amphibians, reptiles and birds, CNS structures including the telencephalon, the diencephalon, the mesencephalon, the rhombencephalon, the myelencephalon and the spinal cord are linearly positioned (Figure 1B). In a human, the extensively developed cerebral cortex overlies not only the basal ganglia, the hippocampus and the amygdala, but also the structures derived from the diencephalon and the mesencephalon. Thus, the linear positioning of CNS structures is apparently unclear (Figure 1B, 2). The cerebral cortex is divided to the frontal, the parietal, the temporal and the occipital lobes (Figure 2A). The primary visual cortex, the primary auditory cortex, the primary somatosensory cortex and the primary motor cortices are located in the occipital, the temporal, the parietal and the frontal lobes, respectively, whose functions are explained in following sections.

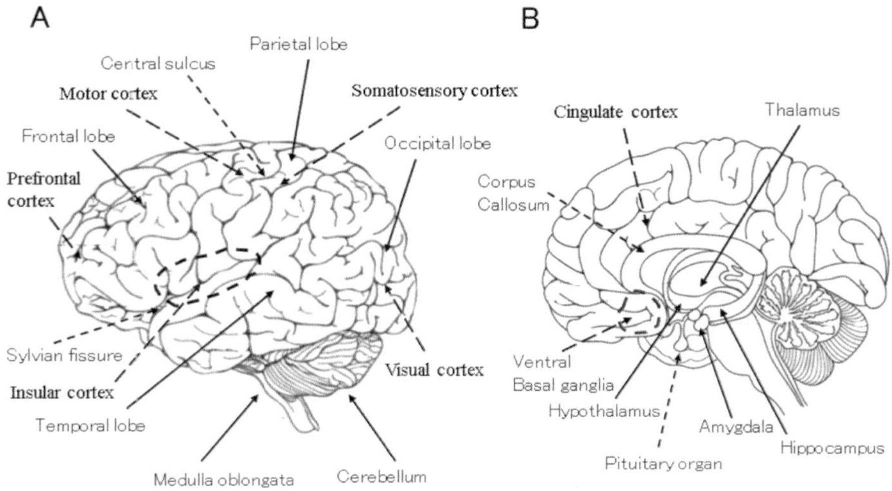

Figure 2. Human brain. A shows the outside view and B shows the inside view of hemisected brain. The cerebral cortex has a number of sulci, making its surface area large. The cerebral cortex is divided to the frontal, the parietal, the temporal and the occipital lobes. The primary visual cortex, the primary auditory cortex, the primary somatosensory cortex and the primary motor cortex are located in the occipital, the temporal, the parietal and the frontal lobes, respectively. The central sulcus separates the frontal and the parietal lobes, and Sylvian fissure separates the temporal lobe from the frontal and the parietal lobes. The thalamus, the hypothalamus, the basal ganglia, the hippocampus, the amygdala, etc. are buried inside of the cerebral cortices. The corpus callosum is a bundle of axons connecting the right and left cerebral hemispheres (modified from Hirano 2019, Purves et al. 2019).

Simple Neuronal Pathways for Reflex

The basal and original role of the nervous system seems to be to transmit information from the sensory organs to the motor systems, so that adequate responses to external stimuli take place. Such a basal function of the nervous system is represented in the control of reflex. Stretch reflexes including the patellar tendon reflex occurs when a muscle is stretched. The extension of a muscle is detected by the muscle spindle, which is a sensory cell buried in the muscle, and the stretch information is sent to the spinal cord through a sensory axon whose cell body is located in a dorsal root ganglion (Figure 3). In the spinal cord, the sensory axon forms synapses directly on a motor neuron controlling the same muscle, from which the sensory stretch signal comes. Thus, only two types of sequentially connected neurons are engaged in the stretch reflex, because of which a stretched muscle contracts.

There are various types of reflexes. If you touch a hot material with your hand, the hand is quickly withdrawn. This reaction is called flexion reflex, and three types of neurons are involved. Information about the high temperature is conveyed to the spinal cord through sensory axons, and is transmitted to interneurons. The interneurons in turn form excitatory synapses on the motor neurons which control flexion muscles of the upper limb. You may think that the withdrawal of your hand from a hot material is caused because you felt the heat. But that is not correct. You

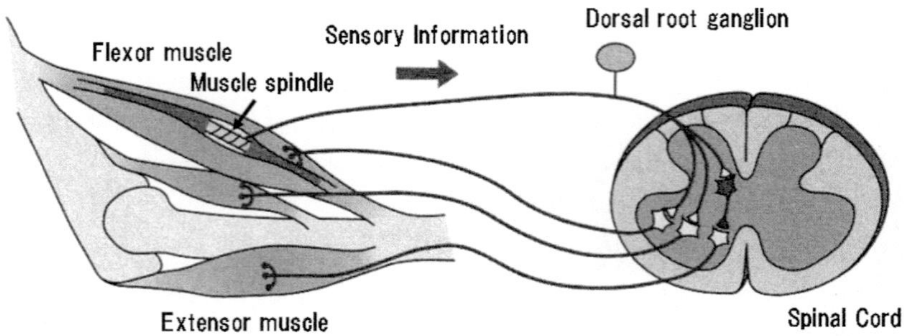

Figure 3. Neuronal pathways controlling the stretch reflex. A sensory axon transmits the information about the length of a flexor muscle from a muscle spindle to motor neurons (white) controlling the same muscle. In addition, a branch of the sensory axon forms synapses on an inhibitory interneuron (dark grey), which in turn suppresses a motor neuron controlling an antagonistic extensor muscle (modified from Hirano 2001, Purves et al. 2019).

withdraw your hand with the spinal reflex before you feel the heat. The hot sensation comes later when the heat information reaches the brain through the spinal cord.

When your head or body is rotated, your eyeballs turn in the opposite direction in order to keep the visual image on the retina stationary. This response is called the vestibulo-ocular reflex (Robinson 2022). The head turn is detected by the semicircular canals in the inner ears, and this information is sent to the vestibular nuclei in the medulla oblongata through the vestibular nerves. Neurons in the vestibular nuclei form synapses on motor neurons controlling extra-ocular muscles. Thus, three neurons pathway control the vestibulo-ocular reflex.

As stated above, neuronal pathways controlling reflexes are simple. They connect sensory cells and muscles with a few neurons. In addition, when a limb is extended by contraction of extensor muscles, the motor neurons controlling antagonistic flexor muscles are suppressed by the inhibitory interneurons in the spinal cord (Figure 3) (Luo 2021). Such antagonistic inhibitions are very common, and presumably support efficient and smooth motor actions. Indeed, rather complicated reflex actions can occur. If you place an acidic substance on the skin on the back of a decerebrated frog, it brushes the substance away by using its leg skillfully. This action is a reflex without any involvement of the brain. Skillful and coordinated actions can be controlled by the spinal cord. I also note that muscles contract not only because of reflexes but also because of motor command signals coming from the cerebral cortex to motor neurons in the spinal cord or the medulla oblongata, which enable voluntary movements.

Sensation and Perception

Most sensory information is not directly sent to motor control systems, but is processed and transformed, so that it becomes more meaningful or useful for an animal to choose an action or to keep the information in the memory. In other words, perception is enabled based on sensation. In the following sections I explain how perception is established through sensory information processing in the brain.

The Peripheral Visual System

In vertebrates, including humans, a visual signal is mainly detected in the retina which is located inside an eyeball at its bottom. The light reaching an eye is refracted by the lens, and focused on the retina. The retina contains not only photoreceptor cells but also pigment cells and several different types of neurons which form a layered structure (Figure 4) (Masland 2012, Luo 2021). Certain visual information processing takes place in the retinal neuronal pathways. Developmentally, the retina is formed from the extension of the neural tube which is called the optic cup (Figure 1A). Therefore, the retina could be regarded as a part of CNS.

There are two types of photoreceptor cells in the retina—rods and cones. Cones are highly concentrated in the central portion of the retina called the fovea, and support high resolution color imaging. On the other hand, rods are found at a relatively low density in the peripheral portion of retina. Rods cannot sense color but are more sensitive to weak light than cones, and serve vision under dim light, where vision is colorless and of a low spatial resolution. Thus, visual qualia are dependent on the characteristics of the photoreceptor cells and/or the retina.

In a rod, the light signal is detected by rhodopsin composed of opsin protein and retinal, which is a small molecule derived from vitamin A. The light changes the conformation of retinal influencing the activity of rhodopsin, which in turn activates the heterotrimeric G protein named transducin. Transducin activates phosphodiesterase, an enzyme that breaks down cyclic guanosine monophosphate (cGMP). On the rod cell membrane, there are cGMP-dependent cation channels, which are opened by the binding of intracellular cGMP. Thus, the decrease in intracellular cGMP concentration caused by the reception of a photon by rhodopsin, results in the decrease in the influx of sodium ions and hyperpolarizes a rod cell. A rod forms synapses on bipolar cells. These synapses secrete glutamate continuously without stimulation, and the hyperpolarization of a rod decreases glutamate secretion. Thus, interestingly, light stimulation does not increase the activity or the output signal of a photoreceptor cell, but rather suppresses the activity using an intracellular signaling cascade.

There are two types of bipolar cells, on-bipolar and off-bipolar cells. Off-bipolar cells increase their activity when light stimulation is off. An off-bipolar cell has an ionotropic glutamate receptor on the postsynaptic membrane facing the photoreceptor cell, and decreases its activity when light is on. On the other hand, on-bipolar cells increase the activity when light stimulation is on. This reversal of visual signal direction in an on-bipolar cell is enabled by the postsynaptic metabotropic glutamate receptor mGluR6, which causes depolarization of an on-bipolar cell through an intracellular signaling cascade, when the amount of glutamate released from a photoreceptor cell is decreased by the reception of the light stimulation.

Bipolar cells also form continuously working glutamatergic synapses on retinal ganglion cells. On-bipolar cells form synapses on on-ganglion cells, whereas off-bipolar cells form synapses on off-ganglion cells. Neither the photoreceptor cells nor the bipolar cells have axons, and both control the secretion of glutamate by the graded change of membrane potential. In contrast, a ganglion cell has an axon which extends to the lateral geniculate body in the thalamus.

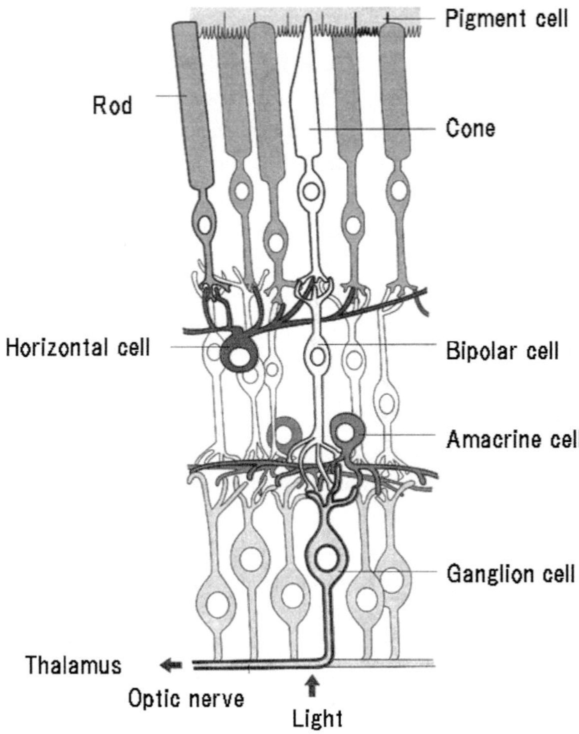

Figure 4. Neuronal pathways in the retina. The light (coming from the bottom), goes through neuronal layers and reaches a photoreceptor cell, either a rod or a cone, located on pigmented cells. Photoreceptor cells form synapses on bipolar cells and horizontal cells. Bipolar cells form synapses on ganglion cells and amacrine cells. Horizontal and amacrine cells are inhibitory interneurons. Ganglion cells send their axons to the thalamus through the optic nerve (modified from Hirano 2021, Purves et al. 2019).

It is to be noted that a photoreceptor cell forms synapses on several bipolar cells, and that a bipolar cell receives synaptic inputs from several photoreceptor cells. In the next step, a bipolar cell forms synapses on several retinal ganglion cells, and a ganglion cell receives synaptic inputs from several bipolar cells. Thus, signal divergence and convergence take place in the retinal neuronal pathways. Therefore, a retinal ganglion cell responds to a light signal reaching a wider area of the retina than a photoreceptor or a bipolar cell does. A photoreceptor cell responds only to the light signal reaching itself.

There are two additional types of neurons in the retina. They are horizontal cells and amacrine cells (Figure 4). Both are inhibitory interneurons, and the former serves in lateral inhibition (Figure 5) (Luo 2021). In lateral inhibition, an excitatory neuron suppresses the activities of neighbor excitatory neurons through the activation of inhibitory interneurons. Lateral inhibition works to enhance the contrast of parallelly transmitted information, because the information transmitted through neighbor neuronal pathways is suppressed. Owing to the lateral inhibitions in the neuronal pathways of the retina, ganglion cells respond most strongly to a bright circle

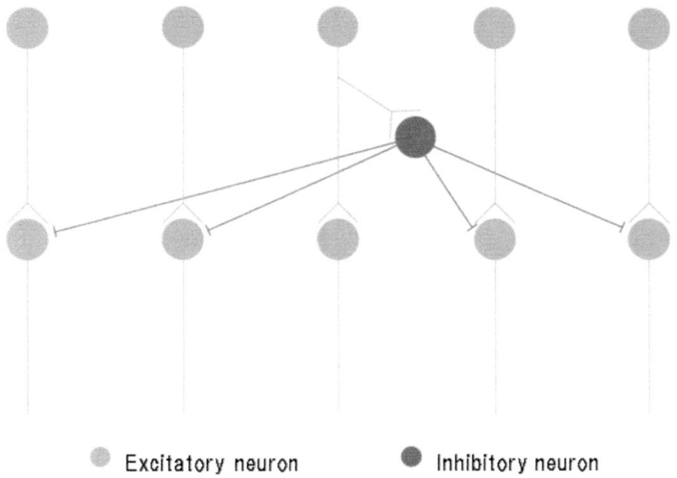

Figure 5. Scheme of lateral inhibition. In parallel neuronal pathways, an excitatory neuron (light grey) form synapses on an inhibitory interneuron (dark grey), which in turn forms inhibitory synapses on excitatory neurons in neighboring pathways. Such a neuronal organization works to enhance the contrast of information transmitted through parallel pathways (modified from Hirano 2019).

surrounded by a dark donut shape, or a dark circle surrounded by a bright donut shape. The sensory signal which activates a neuron is called the receptive field. The receptive field of a ganglion cell is either an on-center off-surround circle or an off-center on-surround circle.

The Central Visual System

Axons of retinal ganglion cells run through the optic nerves and the optic chiasma, and reach the contralateral or ipsilateral lateral geniculate bodies in the thalamus. The neurons there send axons to the posterior part of the cerebral cortex called the primary visual cortex (Figure 6). Some axons of the ganglion cells extend to the superior colliculus in the midbrain. The receptive field of a neuron in the lateral geniculate body is either an on-center off-surround circle or an off-center on-surround circle as those of the ganglion cells, whereas the receptive field of a neuron in the primary visual cortex is an inclined line segment (Figure 7A) (Hubel and Wiesel 1959, Luo 2021). Thus, the receptive fields of cerebral cortical neurons are qualitatively different from those of the retinal ganglion cells or the thalamic neurons.

The primary visual cortex shows a six-layer structure as do most of other cerebral cortical areas (Figure 8). Inputs from the lateral geniculate body reach mainly the stellate cells in layer IV, which sends excitatory outputs to pyramidal neurons in layer II and/or III. Layer II and III pyramidal neurons send excitatory outputs to other cerebral cortical areas related to visual processing, which are called higher order visual cortices. Besides these neurons, there are other types of excitatory neurons in layer V and VI, and various types of inhibitory interneurons in layers II to VI, and they form complicated neuronal pathways including loop circuits. In addition,

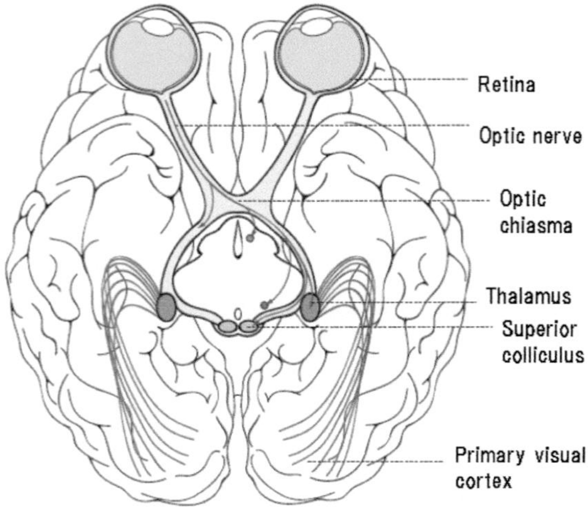

Figure 6. Visual neuronal pathways to the primary cortex. Projection view of horizontal section of the brain is presented. Retinal ganglion cell axons reach the lateral geniculate bodies in the thalamus, whose neurons send axons to the primary visual cortex in the occipital lobe. The visual information in the left field goes to the right thalamus through the optic chiasma. Some retinal ganglion cells send axons to the superior colliculus in the midbrain (modified from Hirano 2019, Purves et al. 2019).

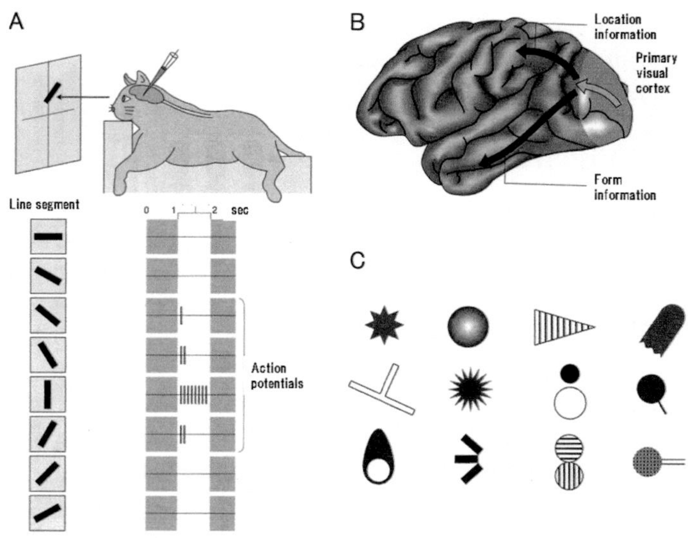

Figure 7. Receptive fields of the visual cortical neurons, and the dorsal and the ventral visual pathways. A, A neuron in the primary cortex of an anesthetized cat responds to a line segment with a specific angle (vertical angle in this example) in a particular location of the whole visual field. B, The dorsal and the ventral pathways for visual information processing. The ventral pathway processes shapes or forms of objects, whereas the dorsal pathway is engaged in stereoscopic vision, perception of movement, etc. C, Higher order neurons in the visual ventral pathway respond to moderately complicated forms such as shown here or faces (modified from Hirano 2001, Hubel and Wiesel 1959, Tanaka 1993, Purves et al. 2019).

the primary visual cortex receives excitatory inputs from other cerebral cortical areas, and sends excitatory outputs to the thalamus and subcortical areas such as the superior colliculus, etc. (Figure 6 and Figure 8).

Higher order visual cortical areas belong to either the dorsal pathway or the ventral pathway (Figure 7B). In the ventral pathway, receptive fields of cortical neurons are moderately complicated shapes (Figure 7C) (Fujita et al. 1992, Tanaka 1993), and their positions do not matter, which are different from those in the primary visual cortex. In the temporal lobe within the ventral visual pathway there are neurons whose receptive fields are faces (Kanwisher and Yovel 2006). Thus, the ventral visual pathway seems to be engaged in the perception of a shape or a form. The processing of visual information for forms seems to make the higher order neurons in the visual system respond not to a fragmental piece of an object but to a somewhat organized shape, such as a face. On the other hand, higher cortical areas in the dorsal visual pathway are engaged in stereoscopic vision, perception of movement, etc. It was reported that a patient with defects in this area cannot perceive the movement of an object, and sees the world as continuous still images (Zihl et al. 1983).

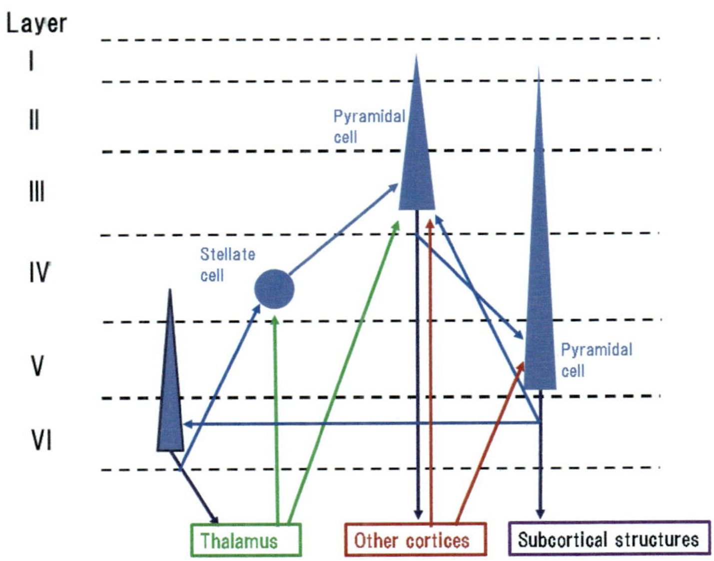

Figure 8. Schematic presentation of layer structure of the primary visual cortex. Inputs from the lateral geniculate body in the thalamus reach mainly stellate cells in layer IV, which send outputs to pyramidal neurons in layer II and III. These pyramidal neurons send outputs to other cerebral cortices such as the higher order visual cortices and the contralateral primary visual cortex. Some thalamic inputs innervate layer II and III pyramidal neurons directly. Layer II, III and V pyramidal neurons receive inputs from other visual cortices including the higher visual cortices and the contralateral primary visual cortex. There are reciprocal connections between layer II, III pyramidal neurons and layer V pyramidal neurons. Layer V pyramidal neurons send outputs to subcortical brain structures such as the superior colliculus and also to layer VI neurons, which send output to the lateral geniculate body (Purves et al. 2019). Besides these excitatory neurons, there are various types of local inhibitory interneurons (not shown here) throughout layers except layer I. Green arrows indicate inputs from the thalamus and red arrows indicate inputs from other cortical areas. Dark blue arrows indicate outputs from the cortex, and light blue arrows indicate intra-cortical connections. Only excitatory inputs and outputs are presented in this figure (Lim et al. 2018).

I would like to note that visual image processing is not linear, straightforward or in one way, but there are multiple loops in the neuronal networks. For example, some higher order cortical neurons send back information to the primary visual cortex, and the layer VI neurons in the primary visual cortex send information back to the lateral geniculate body (Figure 8) (Luo 2021). Thus, the cerebral and thalamic cortical pathways form neuronal circuits with feedback pathways. I also note that there are loop neuronal circuits and various types of inhibitory interneurons within the cerebral cortex (Figure 8). After the above neuronal processing, we recognize a particular person, object, etc. These higher-order perception and/or recognition processes are discussed in chapter five.

In 1909 Brodmann identified forty-seven histologically distinct cerebral cortical regions in a human brain (Loukas et al. 2011). Following studies have increased the number of classified cerebral cortical areas to about two hundreds. Although Brodmann's classification of cerebral cortices was purely anatomical, it has functional correlations. For example, Brodmann's area 4 corresponds to the primary motor cortex, areas 17 and 18 correspond to the primary and the secondary visual cortices, respectively, and areas 44 and 45 correspond to auditory cortices (Amunts and Zilles 2015).

The Auditory System

Sounds, which are actually airborne pressure waves, reach the ears of an animal and cause vibrations in the tympanic membrane located between the outer ear and the middle ear. The vibrations are transmitted to the cochlea in the inner ear via three tiny bones located in the middle ear (Luo 2021). The cochlea is a coiled structure, and contains the organ of Corti, consisting of auditory hair cells, the basilar membrane, tectorial membrane and supporting cells, and also the spiral ganglion, which send auditory information to the cochlea nuclei. Hair cells are sitting on the basilar membrane, which is elastic and is displaced cyclically by sound stimuli, and produces shear force between hair cells and the overlying tectorial membrane. Hair cells in the cochlea have a bundle of stereocilia on their apical end (Fettiplace 2017, Luo 2021). The stereocilia are moved by the shear force, which opens mechanosensitive ion channels located on the stereocilia. Potassium ions come into the cytosol of a hair cell through the mechanosensitive ion channel and depolarize the cell, as the concentration of potassium ions in the extracellular solution facing to the stereocilia is unusually high. The depolarization of a hair cell increases the secretion of neurotransmitter at a synapse formed on the peripheral process of the spiral ganglion neurons, which in turn send excitatory outputs to the cochlea nuclei. By the way, different locations of the basilar membrane along its length are maximally vibrated by different frequencies of sounds. Thus, hair cells located at a particular location of the basilar membrane and its downstream spiral ganglion neurons, respond to a particular frequency of sounds.

The cochlear nuclei neurons send outputs to the inferior colliculus directly and indirectly through the superior olivary nuclei or the lateral lemniscus. Neurons in the inferior colliculus send outputs to the medial geniculate nucleus in the thalamus, which in turn sends the information to the primary auditory cortex in the temporal lobe. Up to the primary auditory cortex, tonotopy which is the spatial arrangement of where sounds of different frequencies are coded, are maintained. I note that the nuclei in the auditory pathways also play roles other than processing of frequencies, such as estimation of the sound source location (Grothe et al. 2010). Some primary auditory cortical neurons respond to various characteristics of sounds, such as a specific temporal interval, the direction of frequency modulation, etc., and some higher order auditory cortical neurons respond to a cry of the same species (Scott 2005, Brefczynski-Lewis and Lewis 2017). In humans some higher order auditory cortical neurons represent specific aspects of speech such as intonation, sentences and speakers (Tang et al. 2017)

The Somatosensory System

We sense body movement, touch, temperature and pain. These sensory signals come from the skin or muscles and go to the spinal cord or the brainstem through axons of sensory neurons, whose cell bodies are located in either the dorsal root ganglia or the trigeminal ganglia (Luo 2021). The morphology of such a sensory neuron is exceptional in having two long axons branched near the cell body (Figure 3). One branch extends to the periphery and another to the spinal cord or the brainstem.

There are various types of sensory neurons in the somatosensory system. The most numerous ones are mechanosensory neurons which convey touch or pain information. They receive mechanical signals at the peripheral nerve endings embedded in some structures in the skin such as hair follicles, Meissner corpuscles and Pacinian corpuscles, or through synapses formed by Merkel cells. Some sensory neurons have peripheral axons with free nerve endings in the skin and detect painful mechanical stimuli, which are called nociceptive neurons. Mechanosensitive ion channels named Piezo play a critical role in mechano-sensation (Coste et al. 2010, Fang et al. 2021). Peripheral endings of proprioceptive neuronal axons are embedded in muscle spindles (Figure 3), tendons or joints, and send information about muscle stretch and tension, which can be related to movements and relative positions of body parts. There are also somatosensory neurons sending information about temperature or chemicals. Many temperature-sensitive neurons also work as nociceptive neurons responding to noxious heat or cold, which cause painful sensations. In the somatosensory system most chemosensory neurons are nociceptive, as they respond to irritants and/or endogenously released chemicals following injury or inflammation. A group of ion channels called TRP (transient receptor potential) channels contribute to chemo-sensation and temperature sensation in the somatosensory system (Caterina et al. 1997, Julius 2013).

Different types of sensory neurons terminate in specific locations in the dorsal part of the spinal cord called the dorsal horn, and form excitatory synapses on distinct types of spinal neurons. These dorsal horn neurons send information to the thalamus relayed by neurons in the medulla oblongata. Then, the thalamic neurons send signals to the primary somatosensory cortex, which is located caudally to the primary motor cortex across the central sulcus in the parietal lobe (Figure 2A). Pain, itch and temperature sensations are primarily conveyed through distinct pathways. Pain information is sent to the hypothalamus and the amygdala, and the latter convey the information to the insular cortex, which is located inside of the cerebral cortex along Sylvian fissure (Figure 2A).

Pain sensation is modulated in the periphery and also by the signal from the brain. The inflammatory response produces bradykinin and prostaglandin, and these molecules enhance the pain sensation through GPCRs and its downstream signaling cascade in the peripheral nerve endings of a nociceptive sensory neuron. The top-down signal from the brain to the pain pathways in the dorsal horn, activate opioid receptors which are GPCRs bound by peptide neuromodulator molecules such as enkephalin (Stein 2016). These opioid receptors are found on both presynaptic terminals and postsynaptic neurons, and suppress the transmitter release and the postsynaptic responsiveness. The suppression mechanism of pain information transmission is explained further in chapter eight, and neuromodulator molecules are explained in chapter nine.

The Olfactory System

Odorants which are molecules that cause olfactory perception, diffuse in the air and go into the nose and get attached to olfactory cilia, which are dendrites of an olfactory receptor neuron located in the olfactory epithelium (Luo 2021). The odorant receptors on olfactory cilia are GPCRs similar to metabotropic neurotransmitter receptors. The binding of an odorant to its receptor activates a type of adenylyl cyclase through heterotrimeric G protein, and cAMP is produced. cAMP binds to a specific nucleotide-gated ion channel, through which sodium and calcium ions come into an olfactory receptor neuron inducing depolarization. The depolarization triggers the generation of an action potential which goes to the olfactory bulb in the brain.

There are more than 1,000 genes encoding odorant receptors in mice, although each olfactory neuron expresses only one type of odorant receptors (Buck and Axel 1991, Mombaerts 1999). Each odorant activates multiple receptor types, and each receptor responds to multiple odorants. Thus, odorants are represented combinatorially in the olfactory epithelium.

Olfactory receptor neurons expressing the same olfactory receptor are distributed broadly in the olfactory epithelium. However, interestingly, these neurons send excitatory axons to the same glomerulus which is a spherical structure in the olfactory bulb (Mombaerts 1999). Thus, olfactory information is represented as a spatial map of glomeruli in the olfactory bulb. Output neurons in the olfactory bulb are mitral cells and tufted cells, which send excitatory outputs to the olfactory cortices (Mori and Sakano 2021). There are multiple inhibitory interneurons in the olfactory bulb, which are suggested to contribute to the lateral inhibition (Luo 2021). The olfactory pathway is a unique sensory system that does not go through the thalamus on the way to the primary sensory cerebral cortex. The outputs of the olfactory bulb are differentially represented in different olfactory cortical areas. A higher order neuron in the olfactory system responds to complicated but meaningful smells such as that of the urine (Courtiol and Wilson 2017).

The Taste System

Chemicals are detected not only by the olfactory receptor neurons but also by the taste receptor cells on the tongue. A cluster of tens of receptor cells form a taste bud having a taste pore at the surface of the tongue (Luo 2021). Taste receptor cells are a kind of epithelial cells, and are innervated by the gustatory nerve coming from the gustatory ganglion. Peripheral axons in the gustatory nerve receive synaptic inputs from taste receptor cells and send outputs to the nucleus of the solitary tract in the brainstem, whose neurons in turn send information to the thalamus. The thalamic neurons send outputs to a part of the insular cortex (Figure 2).

Taste cells are classified into five basic modalities, bitter, sour, sweet, umami and salty. It has been considered that the bitter or sour taste warns an animal that the food might be toxic or spoiled, respectively (Luo 2021). On the other hand, sweet and umami tastes inform an animal that the food is nutritional. The salt taste enables an animal to regulate the sodium ion level in the body fluid. It has been considered that each taste cell responds to one modality in principle.

Sweet, umami and bitter tastes are sensed by GPCRs (Chandrashekar 2000, Nelson et al. 2001). However, there are significant differences between bitter taste receptors and sweet or umami taste receptors. Bitter taste receptors are diverse and there are about thirty types in humans, whereas there are only three types of sweet and umami tastes receptors. In addition, bitter taste receptors respond to a much lower concentration of chemicals than sweet and umami receptors. It has been suggested that the high affinity and divergence of bitter taste receptors help an animal to avoid toxic food effectively (Luo 2021).

Sour and salty tastes are detected by specific ion channels. As for the sour taste a proton ion channel Optopetrin 1 plays a critical role (Tu et al. 2018). In the salt taste sensation, there are two subsystems (Puri and Lee 2021). One responds to only a low concentration of sodium ions and causes appetitive response. In this system the epithelial sodium ion channel plays a critical role in mice, although such an ion channel is not expressed in the human tongue, which suggests that there are other mechanisms in humans. Another salt taste system responds to a high concentration of salt, which activates both the bitter and sour taste receptor cells, and elicits an aversive response (Luo 2021).

It was reported that different parts of the insular cortex are enriched in neurons responding to sweet, umami, bitter and salt tastes (Chen et al. 2011). In addition, it was shown that the artificial activation of neurons in the sweet cortical region using optogenetic techniques, enhanced drinking, and those in the bitter cortical area reduced drinking in mice (Peng et al. 2015). However, it is to be noted that there are also neurons that respond to multiple taste modalities in taste sensory pathways (Luo 2021). By the way, the optogenetic technique is a method to stimulate or inhibit activities of particular neurons in which photo-sensitive ion channels are specifically and artificially expressed, by applying light with a particular wave length. Applications of optogenetic techniques are explained further in chapters six and nine.

The Lower Motor Control System

Skeletal muscles are directly controlled by motor neurons in the spinal cord or in the medulla oblongata, which work as lower motor control systems. A motor neuron receives synaptic inputs from sensory neurons and interneurons in the spinal cord or in the medulla oblongata, and also receives direct or indirect inputs coming from the cerebral cortices. A motor neuron receives not only excitatory glutamatergic inputs, but also inhibitory GABAergic or glycinergic inputs. One example of inhibitory inputs to a motor neuron, controlling a muscle, comes from an interneuron, which receives excitatory inputs from a muscle spindle in an antagonistic muscle (Figure 3). This neuronal pathway is an example of antagonistic inhibition, and works to suppress the simultaneous contraction of extensor and flexor muscles, so that efficient, effective and smooth actions can be performed.

More complicated actions than reflexes can be generated in the lower motor control systems. A decerebrated cat, in which the spinal cord was separated from the upper CNS, could walk on a moving belt conveyer (Shik et al. 1996), indicating that the rhythm and coordinated movements of limbs during walking are generated in the spinal cord. Such a neuronal circuit generating coordinated motor pattern is called a central pattern generator. The existence of a central pattern generator in the spinal cord implies that coordinated and complicated action can be performed without awareness.

Some invertebrate species such as crayfish and squid, and also fish have a few neurons called command neurons (Edwards et al. 1999). The activation of a command neuron induces a coordinated escape reaction involving a number of muscles. In a fish there are a pair of giant command neurons in the hindbrain named Mauthner cells, whose activation causes fast escape swimming (Korn and Faber 2005). Mauthner cells receive sensory inputs from the lateral line organs, etc., and respond to the vibration of water, which could be caused by an approaching predator. As for the decision whether to escape or not is made by the activities of a pair of Mauthner cells in fishes. Thus, activation of only a few command neurons can induce purposeful coordinated behavior.

Inhibitory Neuronal Pathways

As for neuronal pathways including inhibitory neurons and synapses, I have explained antagonistic inhibition and lateral inhibition. The former works around motor neurons and serves to suppress antagonistic muscles, whereas the latter works in sensory systems such as the retina and works to enhance the contrast of sensory signals. They are both important functions of synaptic inhibition, although different roles are played by inhibitory neurons.

The strong and widespread suppression of inhibitory synapses disturbs normal behaviors, and in some cases causes seizure, indicating critical roles of inhibitory neurons in the brain (Treiman 2001). There are important and characteristic inhibitory neuronal pathways in the basal ganglia (Calabresi et al. 2014). Neurons in the internal segment of globus pallidus in the basal ganglia are inhibitory and fire at a high frequency of 50–80 Hz. Most synaptic inputs to these neurons are also inhibitory and come from the striatum in the basal ganglia. This pathway from striatal inhibitory neurons to high-frequency firing inhibitory neurons in the internal segment of globus pallidus, works as a disinhibition pathway, in other words an excitation pathway with serially connected inhibitory neurons. Further explanations of this disinhibition pathway in the basal ganglia are provided in the next section and also in chapter eight. In addition, the importance of excitatory and inhibitory balance in neuronal circuits in the brain are discussed in chapter eleven.

The Higher Motor Control System

The primary motor cortex, which is located anterior to the central groove (Figure 2A), sends motor command signals to the spinal cord directly and indirectly (Lemon 2008). A Betz cell in the layer V of the primary motor cortex is a large pyramidal neuron, possessing a long axon, which directly forms excitatory synapses on motor neurons in the spinal cord (Nolan et al. 2023). Interestingly, the primary motor cortex does not have layer IV. Thus, there are certain regional variations in the architecture of cerebral cortices (Figure 8), implying that there are somewhat different ways of information processing in different areas of the cerebral cortices.

Higher order motor cortices are located anterior to the primary motor cortex in the frontal lobe. They are the premotor areas, the supplementary motor areas and the cingulate motor areas (Figure 2). They send signals to the primary motor cortex and also to the spinal cord. Higher order motor cortices have been considered to engage not in the direct control of motion, but rather in choice, change and preparation of voluntary actions, and a combination of multiple motions (Georgopoulos 1991, Luo 2021). Thus, in motor control systems, higher order motor cortices processing an action plan, etc. send excitatory signals to the primary motor cortex, which sends excitatory outputs to motor neurons, etc. in the medulla oblongata or the spinal cord, and these motor neurons control respective muscles.

The cerebellum and basal ganglia are also involved in motor control (Bostan and Strick 2018). Both of them have direct and indirect connections with motor cortices, and form neuronal pathway loops together with the thalamus. Basal ganglia play an essential role in the initiation of voluntary action (Calabresi et al. 2014), and the cerebellum serves to perform smooth, coordinated movements (Ito 2006), although their functions are multiple and diverse.

In neurological disorders Parkinson's disease and Huntington's disease, the basal ganglia are affected. A Parkinson's disease patient has difficulty in the initiation of movement, whereas a Huntington's disease patient cannot stop excessive movements. Dopamine controls the activities of neurons in the basal ganglia, and in Parkinson's disease the death of dopamine neurons in the substantia nigra pars compacta sending outputs to the striatum takes place. As written in the previous section, there are disinhibition pathways consisting of serially connected GABAergic inhibitory neurons in the basal ganglia. Additional explanations about the basal ganglia are provided in chapter eight, and explanations about Parkinson's and Huntington's diseases are given in chapter eleven.

Cerebellar defects in human patients and experimental animals such as mutant mice show abnormality in coordinated muscle movements called ataxia. The cerebellar cortical circuits have been considered to play critical roles in motor learning which enables smooth coordinated movement using LTD, etc. (Figure 7 of Chapter 3). An example of motor learning is the adaptation of the vestibulo-ocular reflex, which is described in chapters three and six (Ito 2006, Hirano 2013).

Homeostatic Regulation of Body States

Various states of a body such as the blood pressure, breathing conditions, body temperature, nutrition levels, blood osmorality, etc. are sensed and regulated by autonomic nervous systems consisting of sympathetic and parasympathetic systems, the hypothalamus, etc. (Figure 2B) (Luo 2021). Blood pressure and lung inflation are sensed by peripheral terminals of mechanosensitive neurons possessing mechanosensitive ion channel Piezo (Zeng et al. 2018). The increase in blood pressure decreases heart beat rate, and lung inflation inhibits inspiration through the parasympathetic nerves. The blood oxygen level is sensed by sensory cells called glomus cells in the carotid body located near the bifurcation of common carotid artery, and the information is sent to the brainstem. By the way, when an animal senses its predator, the heart beat rate and blood pressure increase because of the activities of sympathetic nerves in preparation for fight or escape, in which epinephrine mediates the signal.

As for food or nutrients, the stomach sends mechanosensory information about its emptiness or fullness, whereas the intestine provides not only mechanical information but also chemical information about nutrients, irritants, etc. Such information reaches the insular cortex, the hypothalamus, etc. (Figure 2). The hypothalamus is a center to regulate various homeostatic processes, and contains neurons that directly sense the body temperature or blood osmorality. The hypothalamus also controls hormonal secretion through the pituitary gland located under it (Figure 2B).

The arcuate nucleus in the ventromedial hypothalamus plays a critical role in regulating the amount of food intake (Figure 2B) (Andermann and Lowell 2017). A rat in which the arcuate nucleus has been lesioned eats more. Following studies clarified that a peptide leptin secreted from adipose tissues binds to its receptor on neurons in the arcuate nucleus. One type of such neurons called POMC neurons are activated by leptin and secrete neuropeptide which reduces food intake (Balthasar et al. 2004). Another type of neurons called AgRP whose activities are inhibited by leptin, release an appetite-stimulating neuropeptide. Optogenetic activation of AgRP neurons promotes eating (Aponte et al. 2011). Interestingly, both POMC and AgRP neurons are modulated by sight or smell before eating, implying that activities of these neurons anticipate the future nutrient level (Luo 2021).

The lamina terminalis in the anterior hypothalamus is involved in the control of blood osmorality. Some neurons there directly sense blood osmorality outside the blood brain barrier and also the level of a peptide hormone angiotensin II, both of which are increased by the dehydration of the animal. Direct activation of such neurons with an optogenetic technique induces drinking (Oka et al. 2015). Additional explanations about the hypothalamus are provided in chapter eight.

Stress Response

Stress is induced by an anticipated threat to well-being, or an actual or anticipated disruption of homeostasis. The sympathetic nervous system provides immediate response to the stress, causing rapid alterations in the animal's physiological state through the regulation of organs in the periphery. For example, when an animal notices a predator, the sympatho-adrenomedullary system rapidly increases the heart beat and the blood pressure by exciting the cardiovascular system. However, such excitation may wane quickly owing to the parasympathetic activation, if the animal recognizes that the predator is not likely to attack it (Ulrich-Lai and Herman 2009). The norepinephrine system including the locus coeruleus in the brainstem is also implicated in stress responses (Schwarz and Luo 2015), which is explained further in chapter nine.

Slower and longer lasting responses to stresses are mediated by the activation of the hypothalamic-pituitary-adrenocortical (HPA) axis through elevations in circulating glucocorticoid hormones. The prime movers of HPA axis are distinct populations of neurons in the paraventricular nucleus of the hypothalamus (Figure 2B), which release corticotropin-releasing peptide hormone into the hypophysial portal plexus of veins. The corticotropin-releasing hormone then travels to the anterior pituitary organ (Figure 2B) and causes the release of another peptide adrenocorticotrophic hormone, which in turn goes to the adrenal cortex in the periphery through blood vessels and causes the increase in cholesterol biosynthesis (Herman et al. 2016). Various steroid hormones including glucocorticoids are produced from the cholesterol, which is further explained in chapter nine.

Glucocorticoids permeate through the cellular membrane, bind to their receptors inside a cell, go into a nucleus, and exert effects through the transcription alteration of gene coding proteins involved in the maintenance of the proper cardiovascular tone, the catabolic actions in the liver, muscle and adipose tissue, and in the regulation of inflammatory and immune responses. Glucocorticoids also induce some actions within seconds or minutes mediated by membrane-localized receptors and intracellular signaling cascades without the involvement of transcription regulation (Nicolaides et al. 2015, Alberts et al. 2023).

Summary and Conclusion

The basal and original role of the nervous system seems to be transmission of information from a sensory organ to a motor system. Such a fundamental function of the nervous system is represented as a simple form in the control of reflex. However, most of sensory information is not directly sent to motor control systems, but processed and transformed so that it becomes more useful for an animal to choose an appropriate response.

In visual information processing, the light signal is detected by photoreceptor cells in the retina, processed in the neuronal pathways with divergence, convergence and lateral inhibition. Then, the retina sends outputs through the axons of ganglion cells to neurons in the lateral geniculate body in the thalamus, which in turn extend axons to the primary visual cortex. There neurons send the information to higher order visual cortices. The receptive field of a neuron in a higher order visual cortex is a moderately complicated shape or a face. Thus, the visual information processing makes higher order neurons to respond to organized forms, etc.

Sounds reach the ears and cause vibrations of the tympanic membrane, which are transmitted to the cochlea and detected by hair cells that send information to the cochlear nuclei. Neurons there send outputs to neurons in the inferior colliculus which project to the medial geniculate nucleus in the thalamus. Then, the thalamic neurons send information to the primary auditory cortex. Up to the primary auditory cortex, the spatial arrangement of neurons responding to respective frequencies of sounds are maintained. In humans some higher order auditory cortical neurons represent specific aspects of speech such as intonation, sentences and speakers.

The binding of an odorant to a G-protein coupled receptor causes depolarization of an olfactory sensory neuron in the olfactory epithelium. Olfactory receptor neurons expressing a same olfactory receptor are distributed broadly in the olfactory epithelium, although these neurons send axons to the same glomerulus in the olfactory bulb. Thus, olfactory information is represented as a spatial map of glomeruli in the olfactory bulb. The neurons of the olfactory bulb send outputs to the olfactory cortices, in which some neurons respond to a smell of the urine.

As explained above, a peripheral receptor cell or neuron responds to an elementary unit of stimulus such as light reaching a particular photo-receptor cell or sound with a particular frequency or a particular chemical substance. In contrast, a neuron in a higher cerebral cortical area responds to a somewhat organized information such as an image of a face, a speaker or a smell of the urine. Thus, processing of sensory inputs in the brain generates neuronal representation of organized and meaningful information which seems to be useful for an animal to make decisions and select actions.

Skeletal muscles are directly controlled by motor neurons in the spinal cord or in the medulla oblongata, which are lower motor control systems. They control not only reflexes but also more complicated motor patterns such as locomotion. Spinal motor neurons receive direct synaptic inputs from the Betz cell in the primary motor cortex. Higher order motor cortices send signals to the primary motor cortex, and have been considered to engage in choice, change and preparation of voluntary actions, and combination of multiple motions. Basal ganglia and the cerebellum are also involved in motor control, although their functions are not confined to them. Both of them have direct and indirect connections with motor cortices, and form neuronal pathway loops together with the thalamus. Basal ganglia play an essential role in the initiation of a voluntary action, and the cerebellum serves to perform smooth, coordinated movements.

Various states of a body such as the blood pressure, the breathing conditions are sensed and regulated by autonomic nervous systems consisting of the sympathetic and the parasympathetic systems and the hypothalamus. The hypothalamus is a center to regulate various homeostatic processes which are performed unconsciously.

References

Alberts, B., K. Hopkin, A. Johnson, D. Morgan, K. Roberts, P. Walter et al. 2023. Essential Cell Biology. 6th Ed. Norton, New York, NY.

Amunts, K. and K. Zilles. 2015. Architectonic mapping of the human brain beyond Brodmann. Neuron 88: 1086–1107.

Andermann, M.L. and B.B. Lowell. 2017. Toward a wiring diagram understanding of appetite control. Neuron 95: 757–778.

Aponte, Y., D. Atasoy and S.M. Sternson. 2011. AGRP neurons are sufficient to orchestrate feeding behavior rapidly and without training. Nat. Neurosci. 14: 351–355.

Balthasar, N., R. Coppari, J. McMinn, S.M. Liu, C.E. Lee, V. Tang et al. 2004. Leptin receptor signaling in POMC neurons is required for normal body weight homeostasis. Neuron 42: 983–991.

Bostan, A.C. and P.L. Strick. 2018. The basal ganglia and the cerebellum: nodes in an integrated network. Nat. Rev. Neurosci. 19: 338–350.

Brefczynski-Lewis, J.A. and W. Lewis. 2017. Auditory object perception: A neurobiological model and prospective review. Neuropsychol. 105: 223–242.

Buck, L. and R. Axel. 1991. A novel multigene family may encode odorant receptors: a molecular basis for odor recognition. Cell 65: 175–187.

Calabresi, P., B. Picconi, A. Tozzi, V. Ghiglieri and M. Di Filippo. 2014. Direct and indirect pathways of basal ganglia: a critical reappraisal. Nat. Neurosci. 17: 1022–1030.

Caterina, M.J., M.A. Schumacher, M. Tominaga, T.A. Rosen, J.D. Levine and D. Julius. 1977. The capsaicin receptor: a heat-activated ion channel in the pain pathway. Nat. 389: 816–824.

Chandrashekar, J., K.L. Mueller, M.A. Hoon, E. Adler, L. Feng, W. Guo et al. 2000. T2Rs function as bitter taste receptors. Cell 100: 703–711.

Chen, X., M. Gabitto, Y. Peng, N.J.P. Ryba and C.S. Zuker. 2011. A gustotopic map of taste qualities in the mammalian brain. Sci. 333: 1262–1266.

Coste, B., J. Mathur, M. Schmidt, T.J. Earley, S. Ranade, M.J. Petrus et al. 2010. Piezo1 and Piezo2 are essential components of distinct mechanically activated cation channels. Sci. 330: 55–60.

Courtiol, E. and D.A. Wilson. 2017. The olfactory mosaic: bringing an olfactory network together for odor perception. Perception 46: 320–332.

Edwards, D.H., W.J. Heitler and F.B. Krasne. 1999. Fifty years of a command neuron: the neurobiology of escape behavior in the crayfish. Trends. Neurosci. 22: 153–161.

Fang, X., T. Zhou, J. Xu, Y. Wang, M. Sun, Y. He et al. 2021. Structure, kinetic properties and biological function of mechanosensitive Piezo channels. Cell Biosci. 11: 13.

Fettiplace, R. 2017. Hair cell transduction, tuning, and synaptic transmission in the mammalian cochlea. Compr. Physiol. 7: 1197–1227.

Fujita, I., K. Tanaka, M. Ito and K. Cheng. 1992. Columns for visual features of objects in monkey inferotemporal cortex. Nature 360: 343–346.

Georgopoulos, A.P. 1991. Higher order motor control. Annu. Rev. Neurosci. 14: 361–77.

Grothe, B., M. Pecka and D. McAlpine. 2010. Mechanisms of sound localization in mammals. Physiol. Rev. 90: 983–1012.

Herman, J.P., J.M. McKlveen, S. Ghosal, B. Kopp, A. Wulsin, R. Makinson et al. 2016. Regulation of the hypothalamic-pituitary-adrenocortical stress response. Compr. Physiol. 6: 603–621.

Hirano, T. 2001. Identity of Brain and Mind (Japanese). Tokyo Kagakudoujin Press, Tokyo.

Hirano, T. 2013. Long-term depression and other synaptic plasticity in the cerebellum. Proc. Japan Acad. Series B 89: 183–195.

Hirano, T. 2019. What is the Brain for? Behavioral Selection and Neuroscience in the Era of AI. (Japanese) Kyoto University Press, Kyoto, Japan.

Hirano, T. 2021. What is Myself? Neuroscientific Consideration of Self and Mind (Japanese) Gentousya Press, Tokyo.

Hubel, D.H. and T.N. Wiesel. 1959. Receptive fields of single neurones in the cat's striate cortex. J. Physiol. 148: 574–591.

Ito, M. 2006 Cerebellar circuitry as a neuronal machine. Prog. Neurobiol. 78: 272–303.

Julius, D. 2013. TRP channels and pain. Annu. Rev. Cell Dev. Biol. 29: 355–384.

Kanwisher, N. and G. Yovel. 2006. The fusiform face area: a cortical region specialized for the perception of faces. Phil. Trans. R. Soc. B 361: 2109–2128.

Korn, H. and D.S. Faber. 2005. The Mauthner cell half a century later: a neurobiological model for decision-making? Neuron 47: 13–28.

Lemon, R.N. 2008. Descending pathways in motor control. Annu. Rev. Neurosci. 31: 195–218.

Lim, L., D. Mi, A. Llorca and O. Marin. 2018. Development and functional diversification of cortical interneurons. Neuron 100: 294–313.

Loukas, M., C. Pennell, C. Groat, R.S. Tubbs and A.A. Cohen-Gadol. 2011. Korbinian Brodmann (1868–1918) and his contributions to mapping the cerebral cortex. Neurosurg. 68: 6–11.

Luo, L. 2021. Principles of Neurobiology. 2nd Ed. CRC Press, Boca Raton, FL.

Masland, R.H. 2012. The neuronal organization of the retina. Neuron 76: 266–280.

Mombaerts, P. 1999. Molecular biology of odorant receptors in vertebrates. Annu. Rev. Neurosci. 22: 487–509.

Mori, K. and H. Sakano. 2021. Olfactory circuitry and behavioral decisions. Annu. Rev. Physiol. 83: 231–256.

Nelson, G., M.A. Hoon, J. Chandrashekar, Y. Zhang, N.J. Ryba and C.S. Zuker. 2001. Mammalian sweet taste receptors. Cell 106: 381–390.

Nicolaides, N.C., E. Kyratzi, A. Lamprokostopoulou, G.P. Chrousos and E. Charmandari. 2015. Stress, the stress system and the role of glucocorticoids. Neuroimmunomodulation 22: 6–19.

Nolan, M., C. Scott, P.R. Hof and O. Ansorge. 2023. Betz cells of the primary motor cortex. J. Comp. Neurol. 532: e25567.

Oka, Y., M. Ye and C.S. Zuker. 2015. Thirst driving and suppressing signals encoded by distinct neural populations in the brain. Nat. 520: 349–352.

Peng, Y., S. Gillis-Smith, H. Jin, D. Tränkne, N.J.P. Ryb and C.S. Zuker. 2015. Sweet and bitter taste in the brain of awake behaving animals. Nat. 527: 512–515.

Puri, S. and Y. Lee. 2021. Salt sensation and regulation. Metabolites 11: 175.

Purves, D., G.J. Augustine, D. Fitzpatrick, W.C. Hall, A.S. LaMantia, R.D. Mooney et al. 2019. Neuroscience. 6th Ed. Sinauer, Sunderland, MA.

Robinson, D.A. 2022. Basic framework of the vestibulo-ocular reflex. Prog. Brain Res. 267: 131–153.

Schwarz, L.A. and L. Luo. 2015. Organization of the locus coeruleus-norepinephrine system. Curr. Biol. 25: R1051–R1056.

Scott, S.K. 2005. Auditory processing—speech, space and auditory objects. Curr. Opin. Neurobiol. 15: 197–201.

Shik, M.L., F.V. Severin and G.N. Orlovskii. 1996. Organization of locomotor synergism. (Russian) Biophizyka 11: 659–666.

Stein, C. 2016. Opioid receptors. Annu. Rev. Med. 67: 433–551.

Tanaka, K. 1993. Neuronal mechanisms of object recognition. Sci. 262: 685–688.

Tang, C., L.S. Hamilton and E.F. Chang. 2017. Intonational speech prosody encoding in the human auditory cortex. Sci. 357: 797–801.

Treiman, D.M. 2001. GABAergic mechanisms in epilepsy. Epilepsia 42 Suppl 3: 8–12.

Tu, Y., A.J. Cooper, B. Teng, R.B. Chang, D.J. Artiga, H.N. Turner et al. 2018. An evolutionarily conserved gene family encodes proton-selective ion channels. Sci. 359: 1047–1050.

Ulrich-Lai, Y.M. and J.P. Herman. 2009. Neural regulation of endocrine and autonomic stress responses. Nat. Rev. Neurosci. 397–409.

Zeng, W., K.L. Marshall, S. Min, I. Daou, M.W. Chapleau, F.M. Abboud et al. 2018. PIEZOs mediate neuronal sensing of blood pressure and the baroreceptor reflex. Sci. 362: 464–467.

Zihl, J., D. von Cramon and N. Mai. 1983. Selective disturbance of movement vision after bilateral brain damage. Brain 106: 313–340.

5

The Neuronal Bases of Higher Brain Functions

Higher Brain Functions: Perception, Recognition, Evaluation, Decision Making and Planning

As stated in the previous chapter, visual information is processed so that assemblies of information about elementary sensory units such as bright spots detected by photoreceptors, give rise to organized and somewhat meaningful representations such as faces in neurons of the higher order visual cortices. Similar kinds of information processing take place in other types of sensation. In auditory information processing, a sensory hair cell in the cochlea of the inner ear responds only to a particular frequency of sounds, whereas a higher order neuron in the auditory cerebral cortex responds to a cry of the same species. Similarly, an olfactory sensory neuron in the nose expresses only one type of odorant receptor, and a higher order neuron responds to a complicated but meaningful smell such as that of the urine.

Then, what happens in the next stage? The identification of the source of sensory signals such as a particular person, a particular food, etc. seems to be necessary for further information processing, and their qualia are supposed to be generated. Such perception and/or recognition are likely to take place in the respective higher order sensory cortices, together with other cerebral cortical areas, etc. Before going into further discussion, I would like to clarify the difference between perception and recognition. According to Cambridge Learner's Dictionary, the definition of perception is "the ability to notice something", while the definition of recognition is "the knowledge of something or someone because you have seen or experienced them before". These definitions indicate that recognition is understanding an object at a higher level than perception.

Figures 1A to C shows an example of experiments which serve to identify the brain areas engaged in the recognition of a face or a house (Blake and Logothetis 2002). Here, overlapped images of a face and a house were presented to a research participant and an fMRI was performed. The participant could recognize the face or the house, and the fMRI image was recorded in both cases. When the patient recognized the face, the fMRI signal increased in the temporal lobe belonging to the

DOI: 10.1201/9781003518822-5

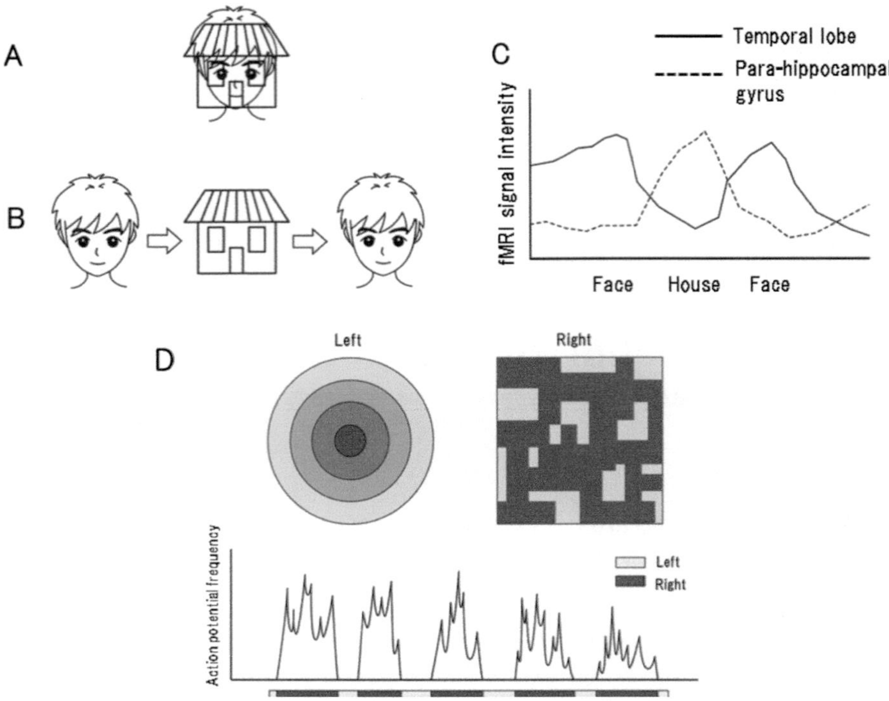

Figure 1. Neuronal activities related to recognition. A research participant saw overlapped images of a face and a house (A), while fMRI was performed. When the participant recognized the face (B), the fMRI signal increased in the visual temporal lobe (C), whereas when the participant recognized the house (B), the fMRI signal increased in the para-hippocampal gyrus (C). (D) Binocular rivalry. Different visual patterns were presented to the left (light grey bar in the bottom) and the right (dark grey bars) eyes of a monkey, which had been trained to press a button when the right object was perceived. A recorded neuron in the visual temporal lobe increased the firing rate of action potentials when the monkey perceived the right object (modified from Hirano 2021, Blake and Logothetis 2002).

higher order ventral visual pathway. On the other hand, when the patient recognized the house, the fMRI signal increased in the para-hippocampal gyrus. Thus, one visual scenery induces different neuronal activities depending on what is recognized; which indicates that neuronal activity underlying recognition is different from that for sensation. More discussion about how neurons code such information is provided in the next section.

The importance, necessity and usefulness of recognized objects should be evaluated and used for decision making or remembrance. These processes are considered to be based not only on the cerebral cortex, but also on the subcortical brain structures such as the basal ganglia, the amygdala, the hippocampus, etc., and emotions influence the process as explained in chapter eight (Hikosaka 2018, Gangopadhyay et al. 2021). Once a decision on what to do is made, one makes an action plan composed of a sequence of behavioral actions, which are carried out in the frontal lobe including the higher order motor cortices, the cerebellum, etc. (Ito 2006), and specific action commands are sent from the primary motor cortex to motor neurons. Some perceived and/or recognized information may not be directly used for immediate choice of behavior, but are stored in the memory for future decision making.

Neuronal Coding

Here, I would like to discuss how a specific object, such as a particular person, an animal, a car, a house, etc. is coded in neurons in the brain. There are two basic ideas. One is that there is a neuron which corresponds to a particular object. This idea or hypothesis is called the cognitive cell theory or a grandmother cell theory, meaning that there is a neuron responding specifically to a grandmother. Another idea is that multiple neurons code a particular object in concert, and that each neuron can participate in the coding of multiple objects. This idea is called population coding, and most researchers, if not all, consider that population coding is working in our brains (Ainsworth et al. 2012). This is because no neuron coding only one specific object has been found, and also there could be actually an infinite number of specific objects and the number of neurons in the brain seems to be insufficient to cover all of them. The number of neurons in the whole brain is estimated to be about 100 billion and that in the cerebral cortex is about 14 billion (Luo 2021).

Within population coding there are two ideas. One is that a relatively small number of neurons having rather strong stimulus preferences code a specific object. This hypothesis is called the sparse coding theory, and is somewhat close to the cognitive cell theory. Another theory is that a large number of neurons having a relatively weak stimulus preference code a specific material, which is called dispersed coding. The high efficiency of sparse coding has been pointed out, and sparse coding has been considered to be prevalent (Spanne and Jörntell 2015, Beyeler et al. 2019).

The assembly of neurons coding a particular object might correspond to NCC, which has been defined as the minimum neuronal mechanisms jointly sufficient for any one specific percept. However, the neuronal assembly coding a particular object could not be confined or identified as a distinct population, and might not be identical in different cases (Shamir 2014). There might be redundancy in neuronal coding as explained in chapter one. By the way, a number of neurons in a functional neuronal assembly is estimated to be 10 to 1,000 (Ainsworth et al. 2012). I also note that the activities of each neuron are quite variable and may be different according to situations. Thus, the simple identification of an assembly of active neurons during perception or recognition of an object without consideration of their activity patterns, interactions and reproducibility, seems insufficient for the rigorous understanding of neuronal mechanisms of perception and/or recognition.

Binding Problem in Recognition

If sparse coding by a limited number of neurons is responsible for the recognition of a particular object with multiple properties, neuronal assemblies coding the object are likely to be composed of a number of smaller neuronal assemblies, each of which codes its restricted part or property. For example, neuronal assemblies coding a grandmother might be composed of sub-assemblies of neurons coding her eye, nose, ear, hand, voice, smell, etc. Similarly, for example neuronal assemblies coding a red large circle could be composed of sub-assemblies coding color red, large size and circular form.

Let's consider that a red large circle and a blue small square exist in the visual field simultaneously. In this condition, neuronal assemblies coding the red and blue colors, large and small sizes, and circular and square forms should be active at the same time (Figure 2). Then, how can one know that a red large circle and a blue small square, but not a red small square and a blue large circle are present.? There should be some neuronal mechanisms that bind or correlate neuronal assemblies coding parts or properties of one object such as color red, large size and circular form together. How this is accomplished has not been elucidated and has been called the binding problem (Feldman 2013, Roelfsema 2023).

One idea or answer to the binding problem is that the action potential occurrences synchronize well among neuronal assemblies coding somethings related to a particular object. According to this idea, in a relatively long time such as 1 sec, all neurons coding red and blue colors, large and small sizes, and circle and square forms are active, but in a shorter time such as 10 msec only neurons coding red, large size and circle are active, and in the following 10 msec neurons coding

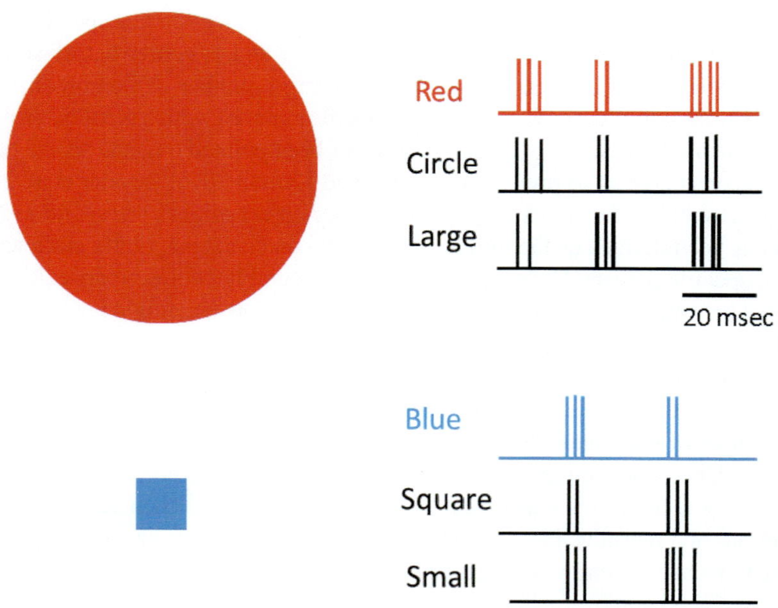

Figure 2. Conceptualistic explanation of a possible neuronal mechanism for the binding. When there are a red large circle and a blue small square in the visual field (left), neurons coding red and blue colors, large and small sizes, and circle and square forms should be active simultaneously. However, in a relatively short time scale, it has been proposed that action potential firing synchronizes well among neurons coding properties of a particular object (right). The vertical lines in the right figures indicate action potentials. Neurons coding red, circle and large size generate action potentials in similar timings (upper 3 traces), whereas those coding blue, square and small size become active in different timings (lower 3 traces). Bursting of action potentials in each neuron may be in the γ frequency range.

blue, small size and square become active, as schematically shown in Figure 2. Synchronized neuronal activities with short delays in distant cerebral cortical areas have been reported (Gregoriou et al. 2009, Ainsworth et al. 2012, Singer 2018). Not only frequencies but also timings of action potential firings have been considered important in neuronal information coding (Ainsworth et al. 2012). However, there is also an idea that neither synchrony nor oscillation of neuronal activities are related to the binding, but that increased neuronal firing rates are crucial for the binding (Roelfsema 2023).

How synchronized neuronal activities are used in information processing and contribute to the formation of the united qualia to a particular object, have not been clarified. An idea or possibility is that synchronized neuronal activities in particular neuronal assemblies, send synchronized synaptic outputs to other assemblies of neurons, resulting in their synchronized activities, and that such sequences of neuronal activities propagate in the brain.

Conceptualization and Words

The united neuronal codes about a particular object should be its neuronal representation. With such a neuronal representation, the object seems to be recognized as a specific existence accompanied with qualia, conceptualized and sometimes expressed with a word in a human. Words and languages are powerful tools used by human beings. They facilitate communication with other people, deep thinking and understanding of complicated matters, events, relations, etc. Words express not only particular objects, but also various matters or affairs including the memory of events or experiences, and also abstract concepts which human beings have created.

It has been reported that in a human, a continuous experience is segmented into the memory as discrete events, and that such a segmented memory is expressed with words and seems to be processed as a single mass in the brain (Baldassano et al. 2017). I discuss more about a kind of short-term memory known as working memory in this chapter, and long-term memory in chapter six.

Sequential and Parallel Information Processing

Both sequential and parallel information processing occurs in the brain. Most of the peripheral sensory systems send information parallelly to the brain. For example, retinal pathways line up parallelly. The information about light stimuli reaching neighboring photoreceptors is sent through adjacent pathways. In the cerebral cortices, visual signals are processed in both the dorsal and the ventral pathways at the same time, although they should be somehow correlated so that we can recognize which object is moving, etc. In any case, parallel information processing is prevalent in the nervous system.

On the other hand, when we speak or write something, we use words sequentially. Thinking with words is sequential, although there might be some background information processing going on in the brain at the same time, which is discussed in the next section. Our voluntary behaviors are also carried out sequentially in principle. It is usually necessary to concentrate on one thing or on one behavior in order to perform meaningful actions without confusion. To complete one particular behavior, one needs to pay attention to matters or objects related to that behavior, and ignore unrelated matters and objects. Thus, choice of information is required in sequential information processing. In the process of information selection, the subjective strength of qualia, emotions and top-down attention seem to play critical roles, which are discussed further in following sections.

Working Memory

Working memory is a kind of short-term memory used for real time thinking, and the frontal lobe of cerebral cortex has been considered to play a critical role in this process (Miller et al. 2018, D'Esposito and Postle 2015). During conversations, words used for a reply are stored in the working memory. In mental arithmetic for summing up a few numbers, the numbers are transiently stored in the working memory. The capacity of the working memory is limited. It has been considered to be about seven, but a report proposed that it might be about four (Cowan 2000).

The limited capacity and transiency of the working memory could be the cause for the sequential nature of our thinking processes, especially when using words. As described in a previous section, our language is sequentially expressed, making parallel thinking difficult. At every moment only a few words or concepts are in the working memory and ready to be used for processing. At the next moment, when a word has been used, another word could get into the working memory and the following information processing or thinking might go on. I think that some background information processing is taking place using related concepts outside of the working memory, and that these concepts might be ready to get into the working memory. However, in the center of conscious thinking there seem to be a limited numbers of concepts or words.

Then, what kinds of neuronal mechanisms underlie the working memory? (D'Esposito and Postle 2015, Meyers 2018). The transient nature of the working memory suggests that the transient activation of neuronal assemblies underlie working memory. If so, and if many neurons are multifunctional and are constituents of multiple neuronal assemblies related to different concepts, the capacity of the working memory is likely to be limited in order to avoid confusion of information processing, which is possibly caused by overlapped activation of neurons belonging to multiple neuronal assemblies. It was reported that some neurons in the prefrontal cortex of a monkey show persistent activities without external stimuli in the waiting period of a particular type of behavioral task named the delayed alternation task (Fuster and Alexander 1971, Kubota and Niki 1971). However, it is to be noted that the prefrontal cortex has not been considered as a site to keep short-term memory, but rather it sends top-down control signals to various brain areas (D'Esposito and Postle 2015). Long-term memory is explained in chapter six.

Rhythm

Many neurons in the central nervous system show various rhythmic activities. The intrinsic property of a single neuron and/or some neuronal circuits can generate rhythmic neuronal activities. Even the refractory period of an action potential, limits the highest firing frequency of a neuron. A simple neuronal loop, consisting of an excitatory neuron and an inhibitory neuron can generate a rhythm. An excitatory neuron can activate an inhibitory neuron, which can in turn suppress the excitatory neuron for a certain period by IPSP (Wang 2010, Bosman et al. 2014, Ainsworth et al. 2012). Rhythmic neuronal activities are generated in central pattern generators underlying locomotion, breathing, etc.

Fast rhythmic activities of multiple neurons can be recorded with electrodes placed on the surface of a head or with an extracellular electrode inserted into the brain. The former reflects activities of many neurons in the large brain area, called the brain wave, and the latter reflects neuronal activities of the restricted brain area and is called the local field potential. The brain waves are classified as δ (1–3 Hz), θ (4–7 Hz), α (8–13 Hz), β (> 14 Hz) waves by frequency. Brain waves change depending on the arousal state of a human or an animal, which is further explained in chapter ten. As for the local field potential, the faster γ wave (around 40 Hz, 26–70 Hz) has attracted much attention from neuroscientists, because the γ wave has been recorded in various brain regions and has been considered to play critical roles in the binding, perception, attention, working memory, etc. (Wang 2010, Bosman et al. 2014, Ainsworth et al. 2012, Singer 2018).

There seem to be generators of the γ wave in the cerebral cortex, in which inhibitory interneurons play critical roles (Wang 2010, Ainsworth et al. 2012, Bosman et al. 2014). The functional roles of the γ wave might be the following. First, it enables information transfer from multiple neuronal assemblies related to different objects at about the same time. Neuronal assemblies representing an object generate action potentials simultaneously at the γ frequency, while other assemblies representing a different object fire also at the γ frequency but with a slight delay (Figure 2). Another role of the γ wave could be to avoid short-term synaptic depression caused by the depletion of the presynaptic synaptic vesicles during the high frequency activation of a presynaptic neuron (Zucker and Regehr 2002).

Loops

There are many information processing loops in the brain and in the whole body including the sensory organs and output organs such as muscle and secretory cells. Such loop structures make information flows bidirectional and complicated. As explained in chapter four, there are loops and mutual interactions among neurons in a local cerebral cortical region. In addition, the higher order visual cortices send feedback information to the primary visual cortex, which sends feedback information to the thalamus (see Figure 8 of Chapter four) (Bruno 2011). Such feedback neuronal connections are prevalent in the brain including the cerebral cortices, the thalamus, the basal ganglia and the cerebellum, which make the neuronal information processes complicated. There are also feedback information flows from the brain to sensory organs. For example, a kind of cochlea hair cell is innervated by an efferent axon from neurons in the medulla oblongata containing acetylcholine, which controls the sensitivity of hair cells to sounds (Fuchs and Lauer 2019).

A role of a feedback is to stabilize the function of a system. If the output is too strong causing an overshooting result, the feedback signal works to suppress the output and keep the system operating in a certain range. Feedback control of the sensitivity to sensory signals has been suggested (Bruno 2011, Fuchs and Lauer 2019). But the roles of the neuronal feedback signals are not likely to be confined to these effects. There are loop structures consisting of excitatory neurons in the cerebral cortex, etc., and such a loop neuronal circuit might serve to retain neuronal activities in the circuit at a high level, which might be possibly involved in the working memory. Neuronal circuit loops containing inhibitory neurons have been proposed to underly the generation of rhythmic neuronal activities as explained in the previous section (Hájos and Paulsen 2009, Wang 2010, Bosman et al. 2014, Ainsworth et al. 2012). Loop neuronal circuits are also prevalent in emotion control systems, which are explained in more detail in chapter nine.

Attention

The definition of "attention" seems to be somewhat ambiguous (Noah and Mangun 2020). The Cambridge Learner's Dictionary defines it as "the state of watching, listening to or thinking about something carefully or with interest". Attention serves to choose sensory information to which one responds (Katsuki and Constantinidis 2014, Noah and Mangun 2020). There are bottom-up and top-down attentions. Bottom-up attention is caused by a salient external stimulus relative to the background. On the other hand, top-down attention is based on prior knowledge, plan and/or current goal. Attention has been considered to enhance activities of neuronal assemblies representing attended objects. It was reported that top-down attention to one of the two superimposed visual images alters fMRI signals, from which an image reflecting the attended object can be reconstructed through decoding using machine learning (Horikawa and Kamitani 2022). Machine learning is explained in chapter seven. The involvement of neuromodulator molecules such as norepinephrine in the control of attention has been reported (Thiele and Bellgrove 2018), which is explained further in chapter nine.

The prefrontal cortex and the posterior parietal cortex have been considered as primary sources of top-down signals in visual attention (Gregoriou et al. 2009). Increased firing rates and synchronization of neuronal activities in the γ-band frequency have been considered to be implicated in the attention processes (Gregoriou 2009, Bosman et al. 2014). The γ-band synchronization was suggested to be a mechanism to label information about an attended object and to distinguish it from unattended ones. The neuronal activity synchronization might serve to bind information about a particular object and to make it stand out with enhanced neuronal activities, which might underly the perception and recognition of a particular object, and the generation of qualia in the subjective mind.

Weighing of Information

For sequential information processing, some contents of sensed and/or stored information need to be selected. Only urgent, critical, important and/or interested matters are selected and processed in the working memory. Other sensed or stored information is neglected. Then, how is this selection performed? In other words, how do we weigh each information? This question is critical to understand our thinking process. Subjectively, we seem to notice salient sensory information such as a loud sound, strong pain, strong smell, etc. which cause bottom-up attention, and also choose objects to which we have paid top-down attention as explained in the previous section. Emotions also play significant roles in weighing of information as explained in chapter eight.

In the objective brain, I think that the neuronal assemblies corresponding to selected objects show high neuronal activities due to either strong sensory inputs or to intense top-down attention signals. Neurons in the activated assembly seem to send strong signals to other neuronal assemblies representing related objects. I think that such information flows and/or interactions between multiple neuronal assemblies may underly our thinking processes, and the wave of neuronal activities may proceed or go around in the brain regions during higher order information processing.

Relationship among Attention, Perception, Recognition, Binding and Working Memory

Attention, perception, recognition, binding and working memory seem to be interrelated. Top-down attention signals are considered to enhance the activities of neuronal assemblies related to the attended object, and the simultaneously activated neuronal assemblies contribute to the binding of the properties of an object, which seems to underly its perception and recognition. The information about such a recognized object by top-down attention is likely to go into the working memory because of its salience. On the other hand, the sensory perception of a salient object is likely to cause bottom-up attention, and to work so that the information about the object enters into the working memory.

One source of a top-down signal has been considered to be the prefrontal cortex, which receives inputs from various brain structures such as other cerebral cortical areas, the thalamus, the amygdala, the hippocampus, etc. (Miller and Cohen 2001, Anastasiades and Carter 2021). The strong inputs from the amygdala suggest that the activities of the prefrontal cortex are significantly influenced by emotions, which is discussed again in chapter eight.

Conflict in Perception and/or Recognition

We usually see an object in the visual field using two eyes simultaneously. Experimentally it is possible to show different objects to each eye of a research participant. In this situation the participant perceives not both objects simultaneously but either object at a time, and the perceived object alternates from time to time. This phenomenon is called binocular rivalry, in which all-or none competition for perception between two objects presented to respective eyes occur, and indicates that perception is distinguished from sensation, which is reception of sensory input signals by the central nervous system.

An experiment on the binocular rivalry using a monkey was performed (Figure 1D) (Blake and Logothetis 2002). Different visual objects were presented to each eye of the monkey, which had been trained to press a button when it perceived one of the two presented objects. Neurons that became active during the button pressing were found in the visual temporal cortex. These neurons were activated not by seeing the objects by itself, but were activated when the monkey perceived one of the two objects. Thus, in the temporal visual cortex there are neurons related not to visual sensation but to perception or recognition.

A number of ambiguous figures have been created as shown in Figure 3. They can be interpreted in more ways than one. When one interpretation is adopted, the recognition of the object based on another interpretation is suppressed. What do you recognize on seeing Figure 3A? You may see two people facing each other. Alternatively, if you pay attention to the black part, you are likely to recognize a vase. What about Figure 3B? You will recognize either a young lady whose face is turned right or the profile of an old woman. In Figure 3C you will see a cube. However, the cube can be interpreted in two ways. One is that the right square is in front and the other is that it is behind. I believe that you cannot recognize any of them in two ways simultaneously. Thus, there seem to be bidirectional inhibitory neuronal pathways between neuronal assemblies representing either of two incompatible recognitions about an ambiguous figure. Such an effect might be brought about by a kind of lateral inhibition in a large scale. In any case, these examples indicate that the visual input of one scenery does not always provide single recognition or interpretation, and that visual recognition is different from sensation.

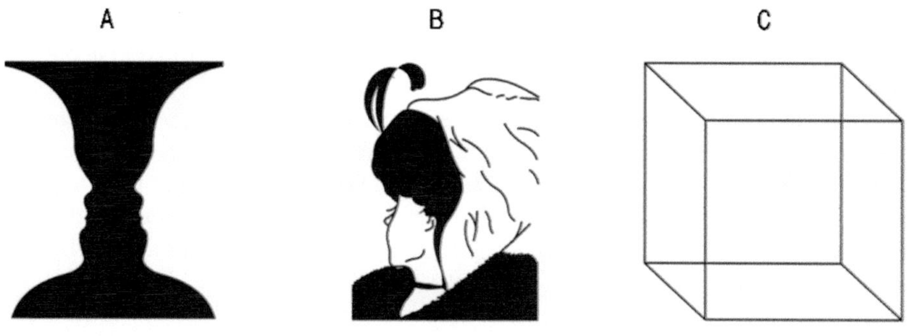

Figure 3. Ambiguous figures. A, The central black part may be recognized as a vase, whereas white area in both sides may be recognized as two people facing each other. This figure is called Rubin's vase (Khalil 2021). B, Either a young lady whose face is turned right or a profile of old woman can be recognized. This figure was created by an unknown person in 19th century, and has been called "wife and mother-in-law", etc. C, A cube which can be interpreted in two ways is shown. One is the right square is in front and the other is it is behind (from Hirano 2021, Blake and Logothetis 2002).

Top-down and Bottom-up Signals

The sensation of pain in a limb is sent to the brain through the dorsal horn of the spinal cord, and this information transfer can be suppressed by the descending signals from the brain as explained in chapter four (Millan 2002). During this pain suppression, peptide neuromodulators such as enkephalin which bind to opioid receptors play important roles. Morphine is an opioid alkaloid substance which binds to opioid receptors. Thus, it has been used as a strong analgesic in pain medication. However, morphine has a number of side effects, among which is that it causes addiction. Pain, neuromodulators and addictions are discussed again in chapters eight and nine.

A primary role of the sensation of pain is supposed to be to inform the brain about an injury and/or an abnormality in a part of the body. Then the animal's brain understands the situation so that it can handle the problem. But if the pain is very strong and the sensation of pain disturbs other information processing by causing a very strong bottom-up attention, it is not good for the animal. For example, when an animal is attacked by a predator and is injured, the animal should run away even if the injury and pain are severe. On such an occasion, pain suppression may help the animal's escape. Such top-down neuronal control on sensory information pathways is present in animals.

Sensation signals originate from peripheral sensory organs and correspond to bottom-up signals. These bottom-up signals are influenced by top-down signals from the brain as explained above. Attention signals from the prefrontal cortex to sensory cortices such as visual cortices also alter neuronal activities. The control of bottom-up signals by top-down signals influences the objective neuronal processing of sensory signals and also of subjective perception and/or recognition (Noah and Mangun 2020, Horikawa and Kamitani 2022). How top-down neuronal signals are generated is a critical question, but has not been elucidated precisely. Presumably, top-down signals reflect weighing of information based on urgency, importance, usefulness, etc. which are judged using memory, future plan, emotions, etc. On the other hand, strong bottom-up sensory signals would cause bottom-up attention, and possibly interrupt ongoing higher order information processing.

Mirror Neurons

It was reported that a neuron in a higher order motor cortex of the frontal lobe which generated action potentials when a monkey grasped something with its hand, also fired action potentials when a human researcher picked up something with a hand in front of the monkey. Following studies have revealed that there are neurons generating action potentials both when an individual performs a certain action and when it observes another individual performing an action with a similar goal. These neurons are named mirror neurons, and it has been suggested that mirror neurons encode not just body movements but rather an action goal, which is defined as an outcome to which the action is directed (Rizzolatti and Sinigaglia 2016).

Mirror neurons have been considered to play roles in imitation, understanding other people's or animal's action, empathy and the theory of mind. How or through what neuronal mechanisms, the activities of mirror neurons are generated is not known. However, the existence of mirror neurons not only in higher order motor cortices but also in other cortical areas including the insular cortex which is related to emotions, suggests that mirror neurons underly various forms of social activities.

Decision Making and Planning

I think that the primary functions of metacognition are making decisions and arranging future action plans based on various recognized and stored information. The decisions and action plans should be aimed to merit an individual, offsprings or a group, or to reduce their demerits. There should be certain conflicts between possible merits and demerits, and/or between short-term and long-term effects. Which merits should be given the first priority are difficult problems. In many cases there are no correct answers, or the best answer cannot be determined because of the above conflicts and also because of the uncertainty about future and present states. Thus, the logical or rational selection of future plans is often difficult, if not impossible. In such cases, the emotions of an individual or his or her experiences might have a significant influence on decision making. Salient qualia, clearly recognized matters and affairs together with emotions seem to be critical in decision making, and they are likely to affect the metacognitive processes.

In the above metacognitive processes, what is happening in the objective neuronal circuits of the brain? Neuronal assemblies corresponding to recognized or memorized objects or concepts in the working memory seem to become active, and these materials are processed sequentially in neuronal circuits using words in some cases. This process might be accompanied with waves of neuronal activities travelling from neuronal assemblies to other neuronal assemblies. At the same time, signals are likely to be sent from neuronal assemblies corresponding to the concepts, objects, matters, etc. in the working memory to other neuronal assemblies representing related concepts, etc. outside of the working memory, resulting in weaker but significant activation of the latter neuronal assemblies. These concepts, etc. are candidates to go into working memory in the next step, and the related neuronal activities and their interactions seem to underly unconscious background thinking. I note that not all the contents in the working memory might be consciously recognized, and that the borders of the working memory might not be so clear. Emotions influence the selection of objects, etc. which are processed in the working memory through strengthening or weakening activities of neuronal assemblies. Salient sensory signals might interrupt ongoing metacognition processes through the related strong neuronal activities and also through inhibitory influences to neuronal assemblies corresponding to other objects, etc. These ideas are speculative, and rigorous verification seems to be difficult at present.

Summary and Conclusion

How is the recognition of an object enabled? One idea is that there is a neuron which corresponds to a particular object, which is called cognitive cell theory. Another idea is that multiple neurons code a particular object in concert and that each neuron participates in the coding of multiple objects, which is called the population coding theory and is widely accepted. Among population coding theory there are two ideas. One is that a relatively small number of neurons having rather strong stimulus preference code a specific object, which is called sparse coding, and is somewhat close to the cognitive cell theory. Another is that a large number of neurons having a relatively weak stimulus preference code a specific material, which is called dispersed coding. The high efficiency of sparse coding has been pointed out. If sparse coding is responsible for the perception of a particular object with multiple characteristics, neuronal assemblies coding the object are likely to be composed of a number of smaller neuronal assemblies, each of which codes its restricted part or property. In this case, there should be some neuronal mechanisms that bind or correlate the smaller neuronal assemblies coding respective parts together. What are the mechanisms? This question is called the binding problem. One idea is that the action potential occurrences synchronize well in a short-time scale among the neuronal assemblies coding a particular object. Synchronized neuronal activities might be used in the following information processing and contribute to the formation of a united qualia to a particular object. Neurons in the central nervous system show synchronous rhythmic activities brought about by local neuronal circuits containing inhibitory neurons, etc. One type of rhythmic activity called γ wave is recorded in various brain regions and considered to play a critical role in the binding, perception, attention, etc.

If different objects are presented to each eye of a research participant, he or she perceives either object at a single moment. This phenomenon is called binocular rivalry, in which all-or none competition for perception between two objects occurs, and indicates that perception or recognition is distinguished from sensation. Binocular rivalry also implies that there are some inhibitory mechanisms that suppress incompatible perception or recognition.

We use words to express materials, concepts, etc., and thinking with words is sequential in principle. In real time thinking, only limited numbers of concepts of around four to nine are processed, presumably to avoid confusion in information processing. The short-term memory used for real time thinking is called the working memory, and transient activations of neuronal assemblies are likely to underlie the working memory. The limitation of the capacity of the working memory necessitates selection of objects, concepts, etc. processed in the working memory. Only urgent, critical, important and/or interested matters are selected and processed. Other sensed or stored information is neglected. Not only the strength of bottom-up sensory signals, but also top-down signals seem to be involved in the selection. Top-down signals include attention and emotion related signals.

The role of metacognition should be making decisions and future action plans based on various recognized and stored information. The decisions and action plans should be aimed to merit an individual, a group or offsprings, or to reduce the demerits. However, there should be certain conflicts between possible merits and demerits, and in many cases, there is no correct answer. In such a case, emotions of the individual, his or her experiences might have a significant influence on decision making. In such processes, neuronal assemblies corresponding to recognized or stored concepts, etc. in the working memory seem to be involved, and they are processed sequentially in principle. Theses processes might be accompanied with waves of neuronal activities travelling from neuronal assemblies to other neuronal assemblies. At the same time, signals are probably sent from neuronal assemblies corresponding to contents in the working memory to other neuronal assemblies representing the related contents, etc. outside the working memory, resulting in weaker but significant activation of those neurons. These contents, etc. are candidates to go into the working memory in the next step, and the related neuronal activities and their interaction seem to underly unconscious background thinking. The overlap of neurons belonging to multiple neuronal assemblies representing different objects, etc. might contribute to the transfer or spreading of information in the neuronal circuits.

References

Ainsworth, M., S. Lee, M.O. Cunningham, R.D. Traub, N.J. Kopell and M.A. Whittington. 2012. Rates and rhythms: a synergistic view of frequency and temporal coding in neuronal networks. Neuron 75: 572–583.

Anastasiades, P.G. and A.G. Carter. 2021. Circuit organization of the rodent medial prefrontal cortex. Trends Neurosci. 44: 550–563.

Baldassano, C., J. Chen, A. Zadbood, J.W. Pillow, U. Hasson and K.A. Norman. 2017. Discovering event structure in continuous narrative perception and memory. Neuron 95: 709–721.

Beyeler, M., E.L. Rounds, K.D. Carlson, N. Dutt and J.L. Krichmar. 2019. Neural correlates of sparse coding and dimensionality reduction. PLoS. Comput. Biol. 15: e1006908.

Blake, R. and N.K. Logothetis. 2002, Visual competition. Nat. Rev. Neurosci. 3: 13–21.

Bosman, C.A., C.S. Lansink and C.M.A. Pennartz. 2014. Functions of gamma-band synchronization in cognition: from single circuits to functional diversity across cortical and subcortical systems. Eur. J. Neurosci. 39: 1982–1999.

Bruno, R.M. 2011. Synchrony in sensation. Curr. Opin. Neurobiol. 21: 701–708.

Cambridge Learner's Dictionary. 4th Ed. 2012 Cambridge Univ. Press, Cambridge, U.K.

Cowan, N. 2000. The magical number 4 in short-term memory: a reconsideration of memory storage capacity. Behav. Brain Sci. 24: 87–185.

D'Esposito, M. and B.R. Postle. 2015. The cognitive neuroscience of working memory. Annu. Rev. Psychol. 66: 115–142.

Feldman, J. 2013. The neural binding problem(s). Cogn. Neurodynamics 7: 1–11.

Fuchs, P.A. and A.M. Lauer. 2019. Efferent inhibition of the cochlea. Cold Spring Harb. Perspect. Med. 9: a033530.

Fuster, J.M. and G.E. Alexander. 1971. Neuron activity related to short-term memory. Sci. 173: 652–654.

Gangopadhyay, P., M. Chawla, O. Dal Monte and S.W.C. Chang. 2021. Prefrontal-amygdala circuits in social decision-making. Nat. Neurosci. 24: 5–18.

Gregoriou, G.G., S.J. Gotts, H. Zhou and R. Desimone. 2009. Long-range neural coupling through synchronization with attention. Prog. Brain Res. 176: 35–45.

Hájos, N. and O. Paulsen. 2009. Network mechanisms of gamma oscillations in the CA3 region of the hippocampus. Neural Netw. 22: 1113–1119.

Hikosaka, O., A. Ghazizadeh, W. Griggs and H. Amita. 2018. Parallel basal ganglia circuits for decision making. J. Neural Transm. 125: 515–529.

Hirano, T. 2021. What is Myself? Neuroscientific Consideration of Self and Mind (Japanese) Gentousya Press, Tokyo.

Horikawa, T. and Y. Kamitani. 2022. Attention modulates neural representation to render reconstructions according to subjective appearance. Commun. Biol. 5: 34.

Ito, M. 2006. Cerebellar circuitry as a neuronal machine. Prog. Neurobiol. 78: 272–303.

Katsuki, F. and C. Constantinidis. 2014. Bottom-up and top-down attention: different processes and overlapping neural systems. Neuroscientist 20: 509–521.

Khalil, E.L. 2021. Why does Rubin's vase differ radically from optical illusions? Framing effects contra cognitive illusions. Front. Psychol. 12: 597758.
Kubota, K. and H. Niki. 1971. Prefrontal cortical unit activity and delayed alternation performance in monkeys. J. Neurophysiol. 34: 337–347.
Luo, L. 2021. Principles of Neurobiology. 2nd Ed. CRC Press, Boca Raton, FL.
Meyers, E.M. 2018. Dynamic population coding and its relationship to working memory. J Neurophysiol 120: 2260–2268.
Millan, M.J. 2002. Descending control of pain. Prog. Neurobiol. 66: 355–474.
Miller, E.K. and J.D. Cohen. 2001. An integrative theory of prefrontal cortex function. Annu. Rev. Neurosci. 24: 167–202.
Miller, E.K., M. Lundqvist and A.M. Bastos. 2018. Working memory 2.0. Neuron 100: 463–475.
Noah, S. and G.R. Mangun. 2020. Recent evidence that attention is necessary, but not sufficient, for conscious perception. Ann. N. Y. Acad. Sci. 1464: 52–63.
Rizzolatti, G. and C. Sinigaglia. 2016. The mirror mechanism: a basic principle of brain function. Nat. Rev. Neurosci. 17: 757–765.
Roelfsema, P.R. 2023. Solving the binding problem: Assemblies form when neurons enhance their firing rate-they don't need to oscillate or synchronize. Neuron 111: 1003–1019.
Shamir, M. 2014. Emerging principles of population coding: in search for the neural code. Curr. Opin. Neurobiol. 25: 140–148.
Singer, W. 2018. Neural oscillations: unavoidable and useful? Euro. J. Neurosci. 48: 2389–2398.
Spanne, A. and H. Jörntell. 2015. Questioning the role of sparse coding in the brain. Trends Neurosci. 38: 417–427.
Thiele, A. and M.A. Bellgrove. 2018. Neuromodulation of attention. Neuron 97: 769–785.
Wang, X. 2010. Neurophysiological and computational principles of cortical rhythms in cognition. Physiol. Rev. 90: 1195–1268.
Zucker, R.S. and W.G. Regehr. 2002. Short-term synaptic plasticity. Annu. Rev. Physiol. 64: 355–405.

6

Learning and Memory

Three Types of Learning

Memory can be defined as a behavioral change caused by an experience, and learning as a process for acquiring memory (Okano et al. 2000, Luo 2021). Learning and memory are crucial for the improvement of decision making and behavioral selection, as they enable an animal to utilize prior experiences, and also consider and evaluate various information about its environment. With more information an animal can make more adequate or better decisions. The curiosity of humans and animals might have evolved to obtain more information about the environment, and to understand it better.

Habituation and sensitization of the gill withdrawal reflex of *Aplysia* as explained in chapter three are simple forms of learning and memory (Castellucci and Kandel 1974, 1976). There are different types of learning and memory, which can be categorized by different criteria such as duration of memory, contents of memory, mechanisms, etc. In the following several sections, I would like to explain three learning paradigms classified by learning processes. They are self-organization, supervised learning and reinforcement learning. Then, I explain different kinds of memory and their neuronal bases.

Self-Organization

During the ontogeny of the brain, receiving sensory signals plays a critical role in the formation of sensory neuronal pathways. It was reported that a cat that had been raised in an environment surrounded only by vertical lines failed to perceive horizontal lines (Figure 1) (Blakemore and Cooper 1970). The primary visual cortex of such a cat was devoid of neurons responding to horizontal lines. Thus, the formation of visual cortical circuits with normal functions, in other words self-organization of neuronal circuits normally processing visual information, requires appropriate visual stimulation during a juvenile period. Similarly, it was also reported that if an eyelid of a cat was sutured for a certain juvenile period, most of primary visual cortical neurons responded only to visual signals from the intact eye but not from the previously sutured eye (Wiesel and Hubel 1963, Espinosa and Stryker 2012). This

DOI: 10.1201/9781003518822-6

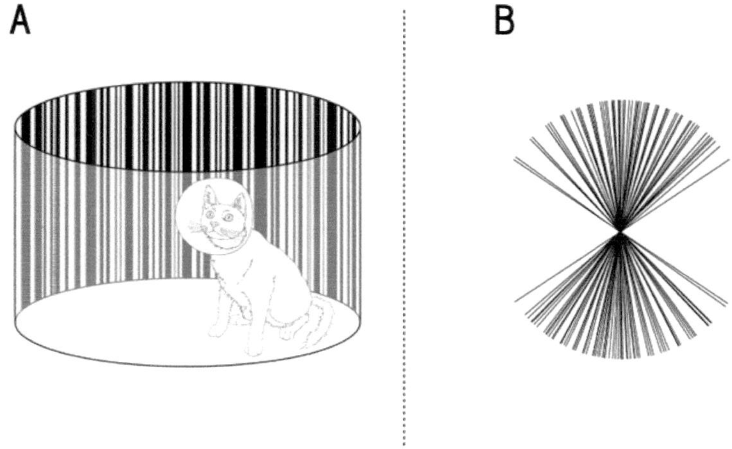

Figure 1. Formation of normal primary visual cortical circuits requires appropriate visual stimulation during a critical period. A: A cat has been raised in an environment surrounded only by vertical lines in a juvenile period. B: In the primary visual cortex of the above cat, most neurons respond to vertical lines and there are few neurons responding to horizontal lines. The lines indicate angles to which neurons in the primary visual cortex responded (from Hirano 2019, Blakemore and Cooper 1970).

phenomenon is called ocular dominance plasticity. The above phenomena found in the formation process of visual cortical pathways might be regarded better as examples of developmental plasticity of neuronal circuits than examples of learning. However, I would like to note that the above plasticity mechanisms depend on experiences and cause behavioral changes, thus they could be regarded as a kind of learning.

Synaptic plasticity mechanisms such as LTP, LTD and STDP which are explained in chapter three have been considered to be involved in the above processes (Espinosa and Stryker 2012). The active and/or effective use of synaptic transmission conveying sensory stimuli causes LTP at the synapses, and establishes effective information transmission. On the other hand, insufficient or ineffective use of synapses causes LTD, and the neuronal information path is eventually weakened or disappears. In self-organization, synaptic plasticity is induced primarily by the activities of synapses undergoing plastic change by themselves, which are different from supervised learning and the reinforcement learning which are explained in following sections. As for the ocular dominance plasticity, the competition between visual signals from each eye seems to be critical (Espinosa and Stryker 2012). The formation and elimination of connections takes place not only in the primary visual cortex but also in other areas of the brain during ontogeny (Sehara and Kawasaki 2011, Kano and Watanabe 2019). I do not think that we are aware of the self-organization processes.

Critical Period and Experience-Dependent Maturation of Neuronal Circuits

Developmental neuronal plasticity such as ocular dominance plasticity occurs in a restricted juvenile period, which is called a critical period (Hensch 2005, Espinosa and Stryker 2012). As for the ocular dominance plasticity of a cat, the critical period is four to eight weeks after the birth. Mechanisms to determine critical periods have been studied, and it was suggested that particular types of inhibitory neurons such as basket cells in the primary visual cortex, play a special role in the initiation of the critical period of ocular dominance plasticity (Fagiolini et al. 2004).

Critical periods have also been reported in other sensory systems, etc. Humans have critical periods for learning languages (Birdsong 2018). Any person can learn any language, if they hear the language during their juvenile periods. But for most of adults learning a foreign language is not so easy. Song birds also have critical periods when they learn the songs (Yazaki-Sugiyama 2019, Rundstrom and Creanza 2021).

Supervised Learning

Supervised learning is a type of learning in which an instructive signal, which has been also called a teacher signal, induces plasticity at a particular synapse. When the outputs of neuronal pathways are correct and/or appropriate, feedback instructive signals are sent to the synapses causing the desired outputs and the synapses undergo potentiation of the transmission. On the other hand, when the outputs are not correct and/or inappropriate, the instructive signals cause depression in the synaptic transmission causing undesirable outputs. An example of the latter case is found in the cerebellar cortical neuronal pathways, which are explained in the next section. Supervised learning has also been used in artificial neuronal circuits with learning ability, which are discussed in chapter seven.

Cerebellum and Motor Learning

In the cerebellar cortex, Purkinje neurons are the sole output neurons extending inhibitory GABAergic axons to the cerebellar nuclei or the vestibular nuclei (Ito 2006). A Purkinje neuron receives a large number of glutamatergic synaptic inputs from parallel fibers which are axons of granule neurons, and a very strong excitatory synaptic input from only one climbing fiber which is an axon of an inferior olivary neuron (Figure 7 of Chapter 3).

Theoreticians proposed that the efficacy of information transmission at synapses between parallel fibers and a Purkinje neuron, changes depending on the activity of the climbing fiber (Marr 1969, Albus 1971, Ito 2001, 2006, Hirano 2013). Among them, Albus (1971) postulated that a granule neuron to a Purkinje neuron synaptic transmission, which has been active in a motor performance and has contributed to cause failure of a motor action, is suppressed depending on the instructive error signal carried by a climbing fiber. Then, Ito and colleagues (1982) reported that coupled activation of parallel fibers and a climbing fiber suppresses a postsynaptic Purkinje neuron activity and its responsiveness to the neurotransmitter glutamate for a long-term. This phenomenon has been called the cerebellar LTD as explained in chapter three, and considered as a cellular basis of motor learning.

An example of motor learning depending on the cerebellar LTD is the adaptation of vestibulo-ocular reflex (VOR) (Ito 1982). VOR occurs as the turn of eyeballs when the head or body is rotated. The head turn is detected by semicircular canals in the inner ears, and this information is sent to motor neurons of extraocular muscles relayed by neurons in the vestibular nuclei, resulting in the turn of eyeballs in the opposite direction of head turn. The purpose of VOR is to keep the visual image on a retina stationary during the head turn. However, the neuronal pathway controlling VOR itself is in one direction without feedback of the visual information. Thus, this pathway itself cannot guarantee that the performance of VOR is adequate. The LTD in a small region of the cerebellar cortex named the flocculus, has been considered to adjust VOR characteristics so that the slip of the image on a retinal is suppressed. The retinal slip information conveyed through a climbing fiber is regarded as an error signal, and works to suppress the parallel fibers to a Purkinje neuron synaptic transmission which has contributed to cause the retinal slip. Purkinje neurons in the flocculus form inhibitory synapses on neurons in the vestibular nuclei, regulate their activities, and thus adjust the dynamic properties of VOR.

Another example of cerebellar LTD-dependent learning is the conditioning of eye blinking (Thompson 2005). If the application of a sound before an air puff stimulation to an eye is repeated in an animal, the eyeblink response begins to take place by the sound application alone. In this conditioning, the sound information is considered to be sent to neurons in the cerebellar nuclei and also to granule neurons, and the mechano-sensory information caused by the air-puff is sent to the motor neurons involved in the eye blinking, and also to an inferior olivary neuron, which is the source of a climbing fiber. Before the application of the sound, the sound information pathway to cerebellar nuclei neurons is proposed to be suppressed by the outputs of the inhibitory Purkinje neurons, which reflect inputs from granule neurons coding the sound signal. When the sound application before the air puff is repeated, it has been proposed that LTD occurs resulting in the suppression of the outputs of the Purkinje neurons, which causes the increase in the activities of neurons in the cerebellar nuclei, which in turn contributes to the occurrence of the conditioned eyeblink response.

A large number of subsequent studies have addressed the relation between the cerebellar LTD and motor learning, and many studies have supported the involvement of LTD in motor learning. However, there are also reports suggesting that motor learning can occur without LTD (Schönewille et al. 2011). Since the discovery of LTD, different forms of synaptic plasticity at other synapses in the cerebellar cortex have been reported, and the contribution of multiple types of cerebellar synaptic plasticity to motor learning has been proposed (Hirano 2013). I think that we are usually unaware of motor learning processes such as improvement of VOR performance, although we may be aware of failures in motor performance.

Reinforcement Learning

Reinforcement learning aims to maximize cumulative long-term rewards in a dynamic environment by balancing the exploration of trials and the exploitation of shorter-term rewards. Reinforcement learning depends on the feedback reward signal, but differs from supervised learning which requires particular input/output pairs (Luo 2021). Reinforcement learning paradigms have been widely used in machine learning in artificial intelligence.

The trial and error method of learning by animals, corresponds to reinforcement learning. A rat can learn that by pressing a specific button among multiple buttons, it can get food. In an imaginary experiment shown in Figure 2A, if a thirsty rat presses

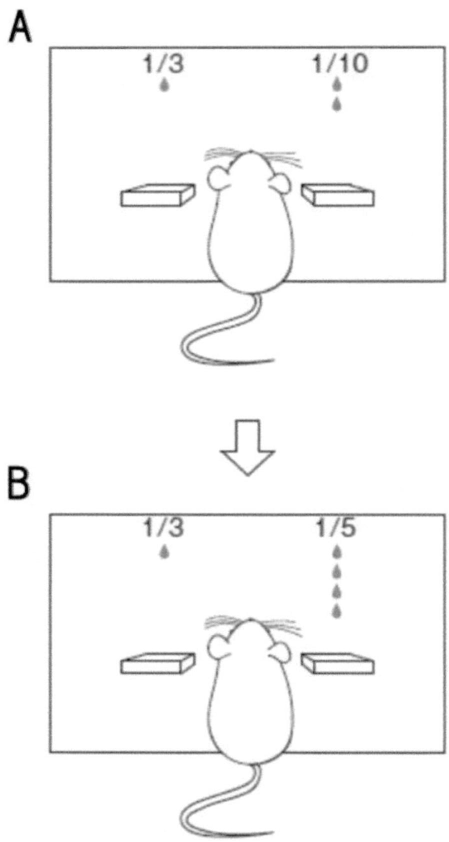

Figure 2. Schematic explanation of an imaginary experiment on reinforcement learning and its conversion. A: A thirsty rat is trained in a condition that it can get one drop of water randomly in about one-third of the cases if it presses the left button, whereas it can get two drops of water by pressing the right button in about one-tenth of the cases. B: After the rat has begun to choose the left button, the condition is changed so that pushing the right button results in the delivery of four drops of water in about one-fifth of the cases without changing the effect of pushing the left button. The rat can get more water drops by pressing the left button in the first condition (A), but in the second condition it can get more by pushing the right button (B) (from Hirano 2019).

the left button, it can get one drop of water randomly once in three times, whereas it can get two drops of water by pressing the right button once in ten times. Here, a rat can get more water drops by pressing the left button, because it can get about ten drops by pressing the left button thirty times, and about six drops by pressing the right one thirty times. In this case, the reward delivery is uncertain or probabilistic as it is given randomly, nevertheless a rat could begin to preferentially choose the left button. Bayesian estimation, which is explained further in the following section, provides a clue about the choice of a button.

Aminergic neurotransmitter dopamine has been considered to convey reward signals in the brain and to enhance synaptic plasticity (Wise 2004, Lerner et al. 2021, Speranza et al. 2021). A dopamine neuron in the ventral tegmental area in the midbrain generates action potentials when an animal gets food. Then, sound stimulation is applied before giving food. After repetitive coupling of the sound stimulation and food delivery, a dopamine neuron comes to respond not to the food delivery but to the sound stimulation (Schultz et al. 1997). This result suggests that the anticipation of getting a real reward can become an internal reward for an animal. This neuronal mechanism might support the behavior or decision making for a possible future reward, and contribute to maximization of the efficacy of getting rewards for a long term. It has been proposed that a reward-related dopamine neuron actually codes a reward prediction error signal (Lerner et al. 2021). If so, reward delivery in an unexpected condition might induce large amounts of dopamine release. Such a mechanism might underly the preferential choice of gambling behavior. I think that we and animals are likely to be aware of reinforcement learning processes, because getting a reward induces emotions such as pleasure. Emotions and dopamine are explained further in chapter eight and nine, respectively.

Bayesian Estimation

Bayesian estimation is a statistical method of calculating the probability of the cause of a particular event, based on results of trials that attempted to cause the event (O'Reilly et al. 2012). It is useful to maximize the expected value of a reward or to minimize the expected value of a loss with uncertain information. In the case of the rat's choice of the right or left button described in the previous section (Figure 2), pushing a button is a candidate of the cause of delivery of water drops, which is the resultant event. By repeatedly pushing the left or the right button and knowing how many water drops are delivered in each case, the rat gets more and more information about the relationship between pushing each button and the delivery of water drops, and the Bayesian estimation becomes more accurate.

Efficient Learning vs Flexibility

A fast learner usually has advantage against a rival in a competition because of the learner's ability to adapt efficiently. When some changes occur in the animal's situation and/or environment, it might be better to alter learned behavior. In the imaginary experiment described in Figure 2, the condition to provide water drops to a rat can be changed. For example, pushing the right button can be changed to cause the delivery of four drops of water in about one-fifth of the cases instead of two in one-tenth cases, without changing the effect of pushing the left button (one drop of water in about one-third cases) (Figure 2B). In this new setting, a rat can get more water drops by pushing the right button, and this should change the behavior from pushing the left button to the right one.

To change learned behavior, one needs to try supposedly unsuccessful or inefficient choices. Thus, occasional attempts to conduct uncertain or probabilistically inefficient behavioral choices, seem to have some meanings and merits in certain occasions. Gambling with an uncertain choice itself might provide an internal reward or pleasure to an animal or a human. Interestingly, a fast learner is not always good at changing learned behaviors. It was reported that a type of genetically engineered mice showing fast learning ability, had difficulty in changing learned behavior and were less flexible (Naude et al. 2016).

Reinforcement learning aims to maximize cumulative long-term rewards in a dynamic environment by balancing exploration of trials and exploitation of shorter-term rewards. Reinforcement learning depends on feedback reward signals, but differs from supervised learning which requires particular input/output pairs. Reinforcement learning of animals underlies trial-and-error learning, and Bayesian estimation seems to be used in reinforcement learning.

Born and Raised

As has been described in this chapter, there are various forms of learning, and we have learned a lot since our birth. Even the neuronal circuit formation is dependent on sensory experiences and resultant synaptic changes. But, formation of basal neuronal structures and circuits such as that of the telencephalon from the neural tube, that of the basic retinal pathways, that of the connection of the retina and the thalamus, etc. are pre- or genetically determined.

Animals have many inborn properties. Herbivores such as horses are born as relatively mature animals, and can stand up immediately after the birth. Rhythmic locomotor movements are generated in the central pattern generator in the spinal cord, which is presumably predetermined at the birth at least in herbivores. A precocial bird such as a duck shows a strong social attachment with a first seen moving object after the birth. There are also many complicated instinctive behaviors whose basal schema seem to be pre-determined. Thus, there are many inborn properties in the brain neuronal circuits. Both inborn and learned properties of the brain are important for its functions, although relative importance seems to vary among animal species. Animals which take a long time to mature, such as humans, may depend more on learning and memory.

Some psychiatric diseases such as schizophrenia, neurodevelopmental disorders have both genetic and environmental risk factors, both of which are considered to affect formation of normally functioning neuronal circuits in the brain (Thapar et al. 2017, Jauhar et al. 2022), which are explained further in chapter eleven.

Declarative and Non-declarative Memory

Memory has been classified into short-term and long-term memory. Working memory, as explained in chapter five, is a type of short-term memory. I note that the term of memory cannot be classified as short and long distinctively, but is rather continuous. There are relatively long-term and very long-term memories, and there are very short-term and relatively short-term memories too.

Among long-term memories, declarative memory and non-declarative memory have been classified (Squire and Dede 2015). Declarative or explicit memory is the memory which can be expressed verbally in humans, and consists of the memory of events called episodic memory, and memory of objects, concepts, facts, etc. which are called semantic memory. On the other hand, non-declarative or implicit memory consists of habits, priming, skills, some conditioned reflexes, etc. They depend on different regions of the brain and are basically independent of each other. The disfunction of either of them does not affect the other in general.

Some characteristics of declarative memory became clear by the studies on a patient, named H.M. (Milner 1972, Squire and Wixted 2011, Luo 2021). H.M.'s medial temporal lobe, which included much of the hippocampus and the adjacent para-hippocampal gyrus, was bilaterally resected in order to relieve severe epilepsy. After the surgery H.M. showed profound impairment of declarative memory. H.M. did not lose general intellectual abilities and also retained some old memories. H.M.'s primary defect was the inability to form new long-lasting memories, which is called anterograde amnesia. H.M.'s case suggests that the hippocampus is critical for the consolidation of memory, and that long-term declarative memory can be stored elsewhere. In addition, H.M. was capable of learning mirror drawing which is a kind of skill, without episodic memory of having practiced the task, supporting that non-declarative memory is independent of declarative memory. Epilepsy is explained in chapter eleven.

How should we consider the declarative memory of animals other than humans? Although they do not use words, animals are likely to have memory about episodes and also about facts such as the memory of their family, etc. Memory about places have been widely used in animal experiments on the hippocampus-dependent declarative memory, as there are neurons responding to a particular place, etc. in the hippocampus (Moser et al. 2008). We and animals seem to be aware of some contents of declarative memory, but there seem to be unnoticed contents as well.

We are usually unaware of non-declarative memory including memory of skill improved through motor learning. Previously described mirror drawing, adaptation of vestibulo-ocular reflex and conditioning of eye blink response are examples of implicit non-declarative memory. They are also called procedural memory, and the cerebellum plays critical roles for their establishment.

A friend of mine previously told me that he had very good visual memory when he was young. He could remember a whole page of a textbook as if it had been photographed without understanding its contents. In an examination he could remember a whole page on which materials related to the question were explained. But to answer the question he needed to read the written sentences on a remembered page, because he had not understood the contents. I myself cannot imagine such a situation, as I have never had such good visual memory. However, if my friend's claim is true, I suppose that such visual memory including unattended materials is formed unconsciously.

Synaptic plasticity has been considered to underly memory formation. For example, LTP in the hippocampus has been considered as a primary cellular basis of the declarative memory formation, and LTD in the cerebellar cortex as a basis of motor learning. Synaptic plasticity in other brain regions such as the cerebral cortex, the basal ganglia, etc. are also likely to contribute to learning and memory. Some long-term memory has been considered to be stored in the cerebral cortical regions related to the property or the attribute of memory (Luo 2021), and I suppose that the stored content is recalled by synaptic inputs which activate the neuronal assembly representing the memorized content.

Engram

"Engram" is the term that was used by Richard Semon a century ago for a physical substrate of memory that is formed by an experience (Josselyn and Tonegawa 2020). Since then, there have been a number of trials to localize or identify engrams. Recent studies on rodents using fear conditioning, and recently developed experimental techniques, such as specific neuronal labelling using a promoter of immediate early gene, optogenetics, and/or transgenic mice, etc. have provided impressive new findings on the engram.

The labelling of neurons which are active during an experience was enabled by using a promoter of an immediate early gene such as c-Fos, Arc or Zif268. Immediate early genes are activated quickly by enhanced neuronal activities (Alberts 2023). The cDNA of a marker protein was fused into a temporally activatable promoter of an immediate early gene and introduced into neurons, which enabled the labelling of neurons that were highly activated during a particular experience. The expression of an immediate early gene can also be detected with immunohistochemistry. Thus, the expression of immediate early genes in a neuron at two different times can be visualized.

These methods were applied to fear conditionings (Reijmers et al. 2007). Two types of fear conditionings have been widely used in neuroscientific research. One is auditory fear conditioning, in which the application of the tone stimulus (conditioned stimulation, CS) is paired with aversive electrical foot shock (unconditioned stimulation, US). The latter causes freezing, which is a fear response. The subsequent application of CS by itself elicits freezing. Another type is contextual fear conditioning, in which a foot shock is applied in a particular chamber. After the foot shock application, bringing the mouse to the chamber itself causes a freezing response, which has been interpreted to mean that the mouse remembers the context, in other words a mouse correlates the chamber with the foot shock. Neurons activated during the pairing of CS and US were labelled with a marker LacZ expressed by the c-Fos promoter. Zif 268 was stained using antibodies after the retrieval of the fear memory caused by CS (Reijmers et al. 2007). The number of neurons marked with LacZ increased with the auditory fear conditioning in the amygdala. And doubly stained neurons with LacZ and the antibody against Zif 268 were found more than the chance level in a mouse which experienced the conditioning and the retrieval after three days. It was also shown that the number of reactivated neurons in the lateral amygdala correlated with the occurrence of the freezing during the tone retrieval, but did not correlate with the freezing during the contextual retrieval. These results indicate that there are neurons which are activated by auditory fear conditioning and become active during the retrieval of fear memory in the amygdala.

Loss of function and gain of function studies on the engram neurons have also been performed. As for the former, viruses expressing CREB (calcium/cyclic AMP–responsive element binding protein), which is a transcription factor that increases neuronal excitability (Alberts et al. 2023), were infected to a small and random population of neurons in the lateral amygdala (Han et al. 2009). It was hypothesized that the neurons infected with CREB were biased for inclusion into the engram of auditory fear conditioning based on previous results. The viruses also contained an inducible construct that causes cell-autonomous ablation, enabling the killing of infected neurons. Ablation of CREB overexpressing neurons after the fear conditioning suppressed freezing response to CS, as if the fear memory of the conditioning was erased.

In gain-of-function studies, inducing the retrieval of a memory has been attempted by artificially reactivating engram neurons in the absence of natural retrieval cues (Josselyn and Tonegawa 2020). In one study, channelrhodopsin, which is a light-gated ion channel derived from algae (Deisseroth and Hegemann 2017), was expressed in a population of hippocampal neurons which were activated during the contextual fear conditioning using a promoter of c-Fos. Then, these neurons were optically reactivated in a different context, which elicited freezing. These results show that optogenetic reactivation of hippocampal neurons activated during the contextual fear conditioning is sufficient to induce freezing without natural retrieval cues (Liu et al. 2012). The hippocampus is explained further in chapter eight.

Memory formation using optogenetics without real experience was also reported (Vetere et al. 2019). In this study optogenetic stimulation of a specific olfactory glomerulus (conditioned stimulus) was paired with the optogenetic stimulation of either appetitive or aversive neural pathways (unconditioned stimulus) in mice. After this pairing, mice showed either an attraction or aversion to the real odor, that was coded in the artificially stimulated olfactory glomerulus. These and other gain-of-function studies suggest that artificial activation of a certain neuronal ensemble can elicit a memory-dependent behavior or can form memory (Josselyn and Tonegawa 2020). The olfactory glomerulus is explained in chapter four.

Furthermore, some recent studies reported synaptic engrams that are ensembles of synapses, which are activated during learning processes, and the manipulation of whose activities influences retrieval of memory (Choucry et al. 2024). In one study, a novel synaptic opto-probe, AS-PaRac1 (activated synapse targeting photoactivatable Rac1) was constructed and introduced into neurons in the motor cortex of a mouse. AS-PaRac1 labeled recently potentiated postsynaptic spines specifically, and induced shrinkage of spines by receiving light stimulation. A motor learning task, in which a mouse was put on a rotating rod, induced substantial spine enlargement in a small subset of neurons. The acquired motor memory was disrupted by shrinkage of the enlarged spines induced by light stimulation of AS-PaRac1, indicating involvement a specific synaptic ensemble in a motor learning (Hayashi-Takagi et al. 2015).

The above studies together indicate that specific neuronal and/or synaptic ensembles are activated during learning processes, and their suppression or activation disturbs or elicits a behavioral response which is presumably based on the memory. It is to be noted that tagged neurons in an ensemble may only be a part of whole neurons in the neuronal ensemble, and the destruction of a part could disturb the retrieval and stimulation of a part could activate whole neurons in an ensemble. The memory engram reminds me of a neuronal assembly representing a recognized object or NCC. However, it is to be noted that the engram of fear conditioning seems to be more than a single percept, and also to be a motor command generator. This is because the information content is the notion of the sound or the recognized environment linked to the foot shock or fear, and the activation of the engram can cause freezing behavior. Differences in examined brain areas in engram studies and perception studies should also be considered. In the former, the amygdala and the hippocampus are mainly examined, whereas in the latter, the cerebral cortical areas are mainly examined. The cerebral cortices might have neuronal groups representing each object, etc., and the amygdala, etc. may have neurons that work to relate objects to fear. An interesting idea has been proposed that linking each memory content is performed through the overlap of neurons belonging to the two neuronal ensembles (Choucry et al. 2024). Neurons belonging to multiple neuronal assemblies representing different objects, concepts, etc. might work in the transfer of enhanced activities from neuronal assemblies to other neuronal assemblies, and in the information processing in and around the working memory, etc.

I also note that the time resolution of detecting memory engram using immediately early gene and that of detecting neuronal assemblies with electrophysiological or live-cell fluorescent imaging studies are very different. The latter may be more than 1,000 times faster than the former. The low time resolution in most of the engram studies was recently overcome by the continuous recording of multiple neuronal activities using calcium-ion-sensitive fluorescent proteins (Zhang and Looger 2024) and a head-fixed miniature fluorescent microscope (Grewe et al. 2017). These methods allowed the recording of the activities of the same neurons continuously for several days. Neuronal activities which were related to the auditory fear conditioning, were recorded from the basal and lateral amygdala. The study demonstrated that the fear conditioning induced both up- and down-regulation of activities of individual neurons to CS, making them similar to those caused by US. During extinction training with repetitive CS application without paring with US, the responses of neurons to CS became different from those to US. It was also shown that the neurons responding to CS exhibited day-to-day turnover, implying that neuronal members of an engram are not fixed.

I would like to note a few additional points about fear conditioning and the use of immediate early gene in labelling activated neurons. One is that the fear conditioning is just one time learning, which is different from most of other types of learning, which require repetition of experiences. This one-time learning is presumably brought about by critical importance of life-threatening fear caused by the electrical foot shock. Another point is that fear conditioning has not been classified into self-organization, supervised learning and reinforcement learning. I also note that the expression of immediately early gene by neuronal activities are not likely to be all-or none. Both over-labelling of neurons with high level basal activities and failure of labelling neurons with relatively low-level activities are likely to occur. These might be parts of reasons for relatively low overlap percentage (10–40%) of doubly tagged neurons by the fear conditioning and the retrieval (Josselyn and Tonegawa 2020). Another reason for the low overlap might be dynamic nature of engrams (Grewe et al. 2017). It is also to be noted that labelling of intrinsically active neurons could be difficult using a promoter of immediately early gene because of their high basal activities.

Summary and Conclusion

Memory can be defined as behavioral change caused by an experience, and learning as a process for acquiring memory. Learning and memory seem to be crucial for the improvement of decision making and behavioral selection, as they enable the utilization of prior experiences. With more information we could make more adequate decision.

The learning paradigms have been classified to self-organization, supervised learning and reinforcement learning. During brain development, receiving appropriate sensory signals plays a critical role in the formation of neuronal circuits which can process sensory information adequately. There are juvenile critical periods for the above self-organizing developmental plasticity mechanisms. Supervised learning is a type of learning in which an instructive signal, which is also called a teacher signal, induces plasticity at a particular synapse. An example is found in the cerebellar cortical neuronal pathways, where cerebellar LTD has been considered to contribute to motor learning. As for the cerebellar LTD, it has been proposed that the granule neuron to Purkinje neuron synaptic transmission which has been active in a motor performance and has contributed to cause failure of a motor action, is suppressed depending on the instructive error signal carried by a climbing fiber. Reinforcement learning aims to maximize cumulative long-term rewards in a dynamic environment by balancing the exploration of trials and the exploitation of shorter-term rewards. It depends on feedback reward signal, but differs from supervised learning which requires particular input/output pairs. Reinforcement learning underlies animals' trial and error learning. Dopamine has been considered to convey reward signals in reinforcement learning, where the anticipation of getting real rewards can become a reward for an animal.

Memory has been classified into short-term and long-term memory, and the latter is further categorized to declarative memory and non-declarative memory. Declarative or explicit memory is the memory which can be expressed verbally in humans, and consists of memory of events called episodic memory and memory of facts, etc. called semantic memory. On the other hand, non-declarative or implicit memory consists of habits, priming, skills, some conditioned reflexes, etc. They depend on different brain regions and are basically independent of each other. Critical roles of the hippocampus for the consolidation of declarative memory and those of the cerebellar LTD for the implicit non-declarative motor skill memory have been proposed. Both inborn and learned properties of the brain are important for its normal functions.

Studies on a physical substrate of memory called engram, have made significant progress using fear conditionings, etc., and recently developed experimental techniques such as specific neuronal labelling using a promoter of immediate early gene, optogenetics, transgenic mice, etc. A neuronal labelling study demonstrated that neurons activated by fear conditioning become active during the retrieval of fear memory in the amygdala. Loss-of-function and also gain-of-function studies of engram have been performed. These studies have indicated that specific neuronal and/or synaptic ensembles are activated during learning processes, and their suppression or activation disturbs or elicits a behavioral response which is presumably based on the memory. The memory engram appears to be somewhat similar to neuronal assembly representing a recognized object, or NCC. However, there are differences. Time resolutions of detected memory engrams have been usually much lower than those of the neuronal assemblies related to perception. In addition, the engram of fear conditioning seems to be more than a single percept, and also a motor command generator. An interesting idea has been proposed that linking of each memory content is performed through the overlap of neurons belonging to two neuronal assemblies.

References

Alberts, B., K. Hopkin, A. Johnson, D. Morgan, K. Roberts, P. Walter et al. 2023. Essential Cell Biology. 6th Ed. Norton, New York, NY.

Albus, J. 1971. A theory of cerebellar function. Math. Biosci. 10: 25–61.

Birdsong, D. 2018. Plasticity, variability and age in second language acquisition and bilingualism. Front. Psychol. 9: 81.

Blakemore, C. and C.F. Cooper. 1970. Development of the brain depends on the visual environment. Nat. 228: 477–478.

Castellucci, V.F. and E.R. Kandel. 1974. A quantal analysis of the synaptic depression underlying habituation of the gill-withdrawal reflex in *Aplysia*. Proc. Nat. Acad. Sci. USA 71: 5004–5008.

Castellucci, V.F. and E.R. Kandel. 1976. Presynaptic facilitation as a mechanism for behavioral sensitization in *Aplysia*. Sci. 194: 1176–1178.

Choucry, A., M. Nomoto and K. Inokuchi. 2024. Engram mechanisms of memory linking and identity. Nat. Rev. Neurosci. 25: 375–392.

Deisseroth, K. and P. Hegemann. 2017. The form and function of channelrhodopsin. Sci. 357: eaan5544.

Espinosa, J.S. and M.P. Stryker. 2012. Development and plasticity of the primary visual cortex. Neuron 75: 230–249.

Fagiolini, M., J.M. Fritschy, K. Löw, H. Möhler, U. Rudolph and T.K. Hensch. 2004. Specific $GABA_A$ circuits for visual cortical plasticity. Sci. 303: 1681–1683.

Grewe, B.F., J. Gründemann, L.J. Kitch, J.A. Lecoq, J.G. Parker3, J.D. Marshall et al. 2017. Neural ensemble dynamics underlying a long-term associative memory. Nat. 543: 670–675.

Han, J.H., S.A. Kushner, A.P. Yiu, H.L. Hsiang, T. Buch, A. Waisman et al. 2009. Selective erasure of a fear memory. Sci. 323: 1492–1496.

Hayashi-Takagi, A., S. Yagishita, M. Nakamura, F. Shirai, Y.I. Wu, A.L. Loshbaugh et al. 2015. Labelling and optical erasure of synaptic memory traces in the motor cortex. Nat. 525: 333–338.

Hensch, T.K. 2005. Critical period plasticity in local cortical circuits. Nat. Rev. Neurosci. 6: 877–888.

Hirano, T. 2013. Long-term depression and other synaptic plasticity in the cerebellum. Proc. Japan Acad. Series B 89: 183–195.
Hirano, T. 2019. What is the Brain for? Behavioral Selection and Neuroscience in the Era of AI (Japanese) Kyoto University Press, Kyoto, Japan.
Ito, M. 1982. Cerebellar control of the vestibulo-ocular reflex-around the flocculus hypothesis. Ann. Rev. Neurosci. 5: 275–296.
Ito, M. 2001. Cerebellar long-term depression: characterization, signal transduction, and functional roles. Physiol. Rev. 81: 1143–1195.
Ito, M. 2006. Cerebellar circuitry as a neuronal machine. Prog. Neurobiol. 78: 272–303.
Ito, M., M. Sakurai and P. Tongroach. 1982. Climbing fibre induced depression of both mossy fibre responsiveness and glutamate sensitivity of cerebellar Purkinje cells. J. Physiol. 324: 113–134.
Jauhar, S., M. Johnstone and P.J. McKenna. 2022. Schizophrenia. Lancet 399: 473–486.
Josselyn, S.A. and S. Tonegawa. 2020. Memory engrams: recalling the past and imagining the future. Sci. 367: eaaw4325.
Kano, M. and T. Watanabe. 2019. Developmental synapse remodeling in the cerebellum and visual thalamus. F1000 Res. 8: F1000 Faculty Rev-1191.
Lerner, T.N., A.L. Holloway and J.L. Seiler. 2021. Dopamine, updated: reward prediction error and beyond. Curr. Opin. Neurobiol. 67: 123–130.
Liu, X., S. Ramirez, P.T. Pang, C.B. Puryear, A. Govindarajan, K. Deisseroth et al. 2012. Optogenetic stimulation of a hippocampal engram activates fear memory recall. Nat. 484: 381–385.
Luo, L. 2021. Principles of Neurobiology. 2nd Ed. CRC Press, Boca Raton, FL.
Marr, D. 1969. A theory of cerebellar cortex. J. Physiol. 202: 437–470.
Milner, B. 1972. Disorders of learning and memory after temporal lobe lesions in man. Clin. Neurosurg. 19: 421–466.
Moser, E.I., E. Kropff and M.B. Moser. 2008. Place cells, grid cells, and the brain's spatial representation system. Annu. Rev. Neurosci. 231: 69–89.
Naude, J., S. Tolu, M. Dongelmans, N. Torquet, S. Valverde, G. Rodriguez et al. 2016. Nicotinic receptors in the ventral tegmental area promote uncertainty-seeking. Nat. Neurosci. 19: 471–478.
Okano, H., T. Hirano and E. Balaban. 2000. Learning and memory. Proc. Natl. Acad. Sci. USA 97: 12403–12404.
O'Reilly, J.X., S.J. babdi and T.E.J. Behrens. 2012. How can a Bayesian approach inform neuroscience? Euro. J. Neurosci. 35: 1169–1179.
Reijmers, L.G., B.L. Perkins, N. Matsuo and M. Mayford. 2007. Localization of a stable neural correlate of associative memory. Sci. 317: 1230–1233.
Rundstrom, P. and N. Creanza. 2021. Song learning and plasticity in songbirds. Curr. Opin. Neurobiol. 67: 228–239.
Schultz, W., P. Dayan and P.R. Montague. 1997. A neural substrate of prediction and reward. Sci. 275: 1593–1599.
Schönewille, M., Z. Gao, H.J. Boele, M.F. Veloz, W.E. Amerika, A.A. Simek et al. 2011. Reevaluating the role of LTD in cerebellar motor learning. Neuron 70: 43–50.
Sehara, K. and H. Kawasaki. 2011. Neuronal circuits with whisker-related patterns. Mol. Neurobiol. 43: 155–162.
Speranza, L., U. di Porzio, D. Viggiano, A. de Donato and F. Volpicelli. 2021. Dopamine: the neuromodulator of long-term synaptic plasticity, reward and movement control. Cells 10: 735.
Squire, L.R. and A.J.O. Dede. 2015. Conscious and unconscious memory systems. Cold Spring Harb. Perspect. Biol. 7: a021667.
Squire, L.R. and J.T. Wixted. 2011. The cognitive neuroscience of human memory since H.M. Annu. Rev. Neurosci. 34: 259–88.
Thapar, A., M.M. Cooper and R. FRCPsych. 2017. Neurodevelopmental disorders. Lancet Psychiat. 4: 339–346.
Thompson, R.F. 2005. In search of memory traces. Ann. Rev. Psychol. 56: 1–23.
Vetere, G., L.M. Tran, S. Moberg, P.E. Steadman, L. Restivo, F.G. Morrison et al. 2019. Memory formation in the absence of experience. Nat. Neurosci. 22: 933–940.

Wiesel, T.N. and D.H. Hubel. 1963. Single-cell responses in striate cortex of kittens deprived of vision in one eye. J. Neurophysiol. 26: 1003–1017.

Wise, R.A. 2004. Dopamine, learning and motivation. Nat. Rev. Neurosci. 5: 483–494.

Yazaki-Sugiyama, Y. 2019. Neuronal mechanisms regulating the critical period of sensory experience-dependent song learning. Neurosci. Res. 140: 53–58.

Zhang, Y. and L.L. Looger. 2024. Fast and sensitive GCaMP calcium indicators for neuronal imaging. J. Physiol. 602: 1595–1604.

7

A Comparison of The Brain and A Computer System

Computer Systems and Artificial Intelligence

A personal computer system is composed of the central processing unit (CPU), memories, and peripheral input and output devices such as a keyboard, a mouse, a camera, a scanner, a monitor, a speaker and a printer (Figure 1). Communication with other computers through the internet also works as an input and output system. The CPU controls data processing and performs calculations. The CPU has transient memory, which is called cash memory, that can be processed efficiently within a short time. There are other types of memories in a computer system. These are classified as Random Access Memory (RAM) which works only when the power switch is on, and Read Only Memory (ROM), which is retained even when the power switch is off, such as a hard disc. Thus, a computer system possesses several types of input, output and memory devices, which is similar to the brains of humans and other animals. CPU and memories are composed of integrated circuits using semiconductors, etc., and the processing speed is very high. One step of processing takes about 0.1 nano seconds (nsec) (Patterson and Hennessy 2012).

Artificial neuronal network systems similar to computer systems equipped with learning ability have been developed. A pioneering example is perceptron (Figure 2). It has three layers, input, middle and output layers. The components of the middle layer receive inputs from some of the input layer components and send outputs to all the components of the output layer. The connection between the middle layer components and an output layer component can be enhanced when the latter responds to a desired pattern of inputs. This working scheme of perceptron corresponds to supervised learning explained in chapter six, and perceptron can learn input patterns such as English or numerical characters. In other words, an output layer component can be made to respond to a particular input pattern. The cerebellar cortical circuits (Figure 7 of Chapter 3) have been regarded as a type of natural perceptron (Ito 2006). The mossy fibers coming from the brainstem neurons innervating granule neurons correspond to the input elements, and the granule neurons correspond to the middle layer components, and the Purkinje neurons correspond to the output layer

DOI: 10.1201/9781003518822-7

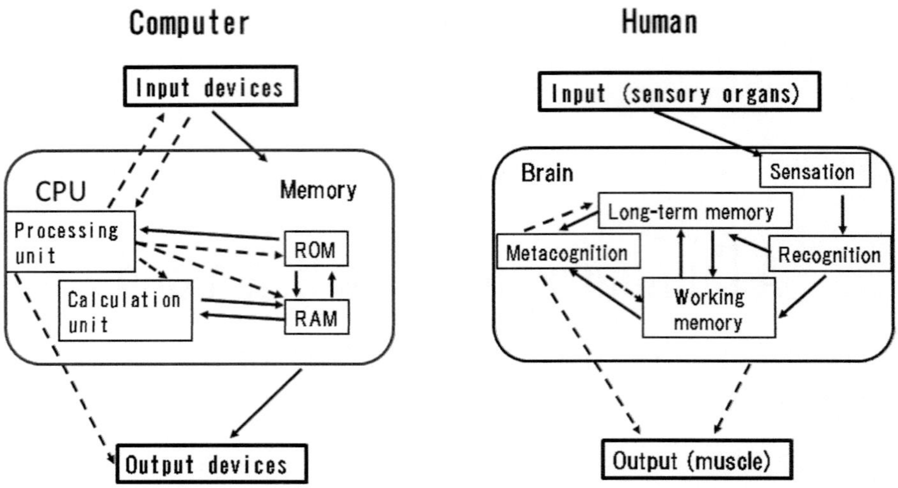

Figure 1. Scheme of a computer system and a brain. Both of them have peripheral input and output devices, processing units and memories. The inputs devices in an animal or a human are sensory organs, whereas they are a camera, a mike, a keyboard, etc. The output devices in an animal are muscles, etc., whereas they are a monitor, a printer, etc. in a computer system. Both of them have short-term and long-term memories. In a computer short-term memory are cash memory (not shown here) and RAM, and long-term memory is ROM. In human short-term memory includes working memory, which plays roles in the center of metacognitive functions. In a computer system, CPU performs calculation and processing of data in the cash memory, and controls memories, input and output devices. Both the brain and a computer system are designed to process information that they get and/or is stored. Solid arrows show information flows, and broken arrows indicate instruction signals. I note that neither all contents nor all information and instruction flows are shown here. For example, strong sensation could directly affect metacognition, and sensory inputs can directly cause motor action by reflexes in an animal or a human.

components. A climbing fiber works as a teacher and sends an instruction signal to a Purkinje neuron, and regulates the synaptic transmission between a granule neuron and a Purkinje neuron through the induction of LTD as explained in chapters three and six, resulting in the formation of an appropriate response pattern of the Purkinje neuron (Ito 2001, Hirano 2013).

By increasing the number of processing layers and data, and adopting a reinforcement learning scheme, artificial intelligence (AI) with a very high learning ability has been developed. AI not only distinguishes various patterns of data but also performs a number of intelligent tasks which include verbal communication, translation, writing sentences, driving a car, etc. Graphic processing units (GPU) have been widely used in AI. GPU is designed to conduct real time processing of images, and can perform parallel calculation of data. AI's extraordinary splendid performance has been enabled by using very fast information processing with semiconductor integrated circuits and huge data sets together with its learning ability.

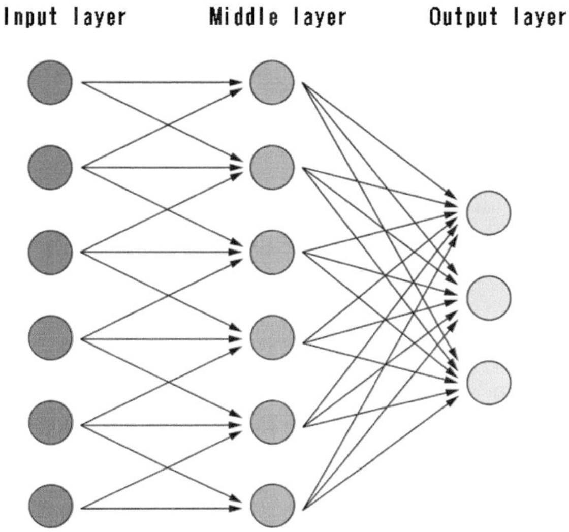

Figure 2. Perceptron. Perceptron consists of input, middle and output layers. Middle layer components receive inputs from some of input layer components and send outputs to all components of the output layer. The connection between a middle layer component and an output layer component can be enhanced when the latter responds to a desired pattern of inputs. Perceptron can learn input patterns such as numerical characters. In other words, an output layer component can be made to respond to a particular input pattern (modified from Hirano 2021).

Similarities of the Brain and a Computer

The overall design of a computer system and the brain are somewhat similar (Figure 1). Both of them have peripheral inputs and outputs devices, processing units and memories. The inputs devices in an animal or a human are sensory organs such as the eyes, the ears, etc., whereas in a computer system, they are a camera, a mike, a keyboard, etc. The output devices in an animal are muscles, secretory organs, etc., whereas they are a monitor, a printer, a speaker, etc. in a computer system.

The central processing unit in a human seems to be the metacognitive system, which performs the evaluation of various information for decision making using memory. The metacognitive system works together with neuronal circuits processing sensory inputs and those controlling motor actions to conduct appropriate responses to stimuli from the environment. Parallel processing of sensory information in an animal and parallel processing of data by a GPU in a computer system may be somewhat analogous. Metacognitive processes and/or conscious thinking in a human use working memory. In a computer system, CPU performs the calculation and processing of data in the cash memory. Working memory in a human and cash memory in a computer are somewhat similar. Both are transient and readily used by a main processing unit. In humans top-down attention contributes to the selection of sored contents which should be processed in the working memory. In a computer system, the CPU controls writing down and readout of memory contents to and from RAM and ROM. Both a human brain and a computer system have different types of memories. Some are transient and are not retained, whereas some are kept for a long time, although access to latter memories is limited, difficult and/or takes a long time. Both the brain and computer systems are designed to process information that they get and/or store.

A computer system processes information digitally. Neuronal coding of information by the frequency of action potentials is also digital, as the generation of the action potential is all-or-none. However, I note that the brain also uses synaptic transmission and neuromodulator molecules, which are analogue systems for information transfer.

Differences between the Brain and a Computer

One critical difference between the brain and a computer system is the speed of processing. One step is about 0.1 nsec in a computer, whereas synaptic transmission needs around 1 msec. The generation of an action potential by the summation of EPSPs requires around 10 msec, and the conduction of an action potential also takes time. Thus, elements of a computer process information more than 10^7 times faster than those of the brain. Faster information processing is advantageous for an animal to survive and to win in competitions with rivals. Therefore, animals have developed several ways to speed up information processing in the brain at the expense of accuracy, which are discussed in the following sections.

Another critical difference are their purposes. As stated in chapter two, the brain has evolved so that animals behave to survive as an individual, and/or to contribute to a group to which it belongs. On the other hand, any computer system or AI has been designed to perform specific purposes that humans set for them. They do not have independent purposes so far, although recent fast development has increased the possibilities of using AI in various tasks such as controlling machines, pattern recognition, communication, presentation of required information and/or candidate solutions of problems. Recent AI, called AI agents, perform tasks without detailed instruction, as if behaving independently apparently. In communications, the distinction of AI from a human might have become difficult in some cases. Though, at present there is no general-purpose independent AI, such as a robot with its own free will. If such a robot is made, what its purpose should be is a difficult problem. Also, whether AI can have mind, consciousness and emotions is an interesting but difficult question to answer. These topics are discussed again in chapter thirteen.

The materials that compose a brain and a computer system are very different. The brain is mainly made from organic materials such as proteins, lipids, etc. and also water and ions, whereas a computer system is made from inorganic materials such as semiconductors, metals, etc. The energy sources are also different: the brain uses chemical energy such as those obtained by the decomposition of ATP, etc. (Alberts et al. 2023), whereas a computer uses electrical energy. An individual brain has a limited lifespan, whereas a computer system does not have a certain lifespan, although it is likely to break down due to deterioration over time. Death or the limited lifespan of an individual animal seems to influence the purpose of its brain.

We should not belittle the fact that we humans are living beings who cannot avoid aging and death, and that we reproduce offsprings by mating; whereas computer systems are different. Computer systems do not have inborn purposes other than those set by humans. On the other hand, living beings are destined to work towards the survival of the species as explained in chapter two. I think that the self, mind and consciousness have evolved a strategy to improve the ability for the survival of the individual, and/or contribute to the survival of the group to which it belongs, which in principle should ultimately contribute to the survival of the species. As it will be explained in chapter eight, emotions seem to have evolved to merit the survival of an individual and its closely related individuals. I note that, as far as I know, no computer system contains an information processing system which resembles the emotion control systems in animals or humans. I think that the implementation of such emotion-like functions to a computer system is difficult and/or inappropriate.

There are also differences between the brain and computer systems in the ways information is processed, such as diffusive transmission by neuromodulators, occurrence of global state changes, and the existence of various and complicated loop structures, including inhibitory pathways in the brain, which are explained further in following sections and in chapters five, eight, nine, ten and eleven.

Diverse and Complicated Neuronal Circuits

In the brain there are various types of neuronal circuits and also various ways of information transmission (Luo 2021). Both parallel and sequential processing take place. Parallel processing is conspicuous in the peripheral sensory nervous systems such as the retina (Masland 2012), whereas metacognitive processes using working memory and words are basically sequential. The brain has various types of neurons with different morphologies and functions, different numbers of inputs and outputs, and different types of basal activities including intrinsic activities (Huguenard and McCormick 1992, Raman and Bean 1999).

Not only excitatory neurons and synapses are present in the brain, but also inhibitory neurons and synapses. Inhibitory neurons and synapses work in several ways such as lateral inhibition, antagonistic inhibition, disinhibition, suppression of less important information to focus on the processing of critical information, etc. Synaptic connections are usually one-to-one hard-wired connections. However, there are another type of neuronal transmission called volume transmission (Özçete et al. 2024). Some neurotransmitter or neuromodulator molecules such as amines are released to wider extracellular spaces from some axon terminals and affect a number of nearby postsynaptic neurons simultaneously.

The effects of neurotransmitter on a postsynaptic neuron are not confined to fast responses such as EPSP and IPSP. There are also slower and longer-lasting effects through GPCRs and intracellular signaling cascades, whose outputs are sometimes neither simple excitation nor inhibition. There are also various types of activity-dependent synaptic plasticity with different induction conditions, different time courses, etc.

I think that the above diversity of neuronal systems has developed so that efficient and/or fast information processing for better decision making can be performed with limited numbers of and relatively slow working speeds of neurons and synapses. Such diversity of information processing machineries and functions are found in neuronal circuits of the brain, but not in computer systems. I also note that our brains show the circadian rhythm and different functional states such as awaking and sleep, in which same inputs yield different outputs through the same neuronal pathways and circuits, which are described in chapter ten.

Efficient Information Processing with Slow Elements in The Brain

As explained, neurons and synapses need substantially more time for information processing compared with integrated circuits in computer systems made from semiconductors, etc. Fast responses and decision making are advantageous in the competitive environment of an animal where a number of rivals and enemies exist, which necessitates the evolution of nervous systems for faster information processing.

Contrast enhancement with lateral inhibition and filling-in the lacking part of an object are likely to be useful for fast recognition, which then contribute to fast decision making and reactions. The suppression of less important information, in order to concentrate on presumably critical objects or matters with attention, would prevent confusion of information processing and decision making, resulting in fast responses. Such various properties of neuronal circuits are designed to accomplish efficient information processing by a limited number of elements with relatively slow working speeds.

Contrast Enhancement

Lateral inhibition works to enhance the contrast of parallelly transmitted information. Neuronal pathways possessing lateral inhibition are found in the retina where a type of inhibitory neurons called horizontal cells play a role (Foley 2019, Luo 2021). An example demonstrating the effect of lateral inhibition is presented in Figure 3A. It may seem that the darkness of the grey in the middle band is graded: the left side is darker than the right side. But, in reality the darkness of the middle band is uniform as shown in the bottom figure. You can confirm it by hiding the upper and lower bands of the top figure. This effect is caused by lateral inhibition. Bright grey on the left side of the upper and lower bands exerts inhibitory effects on the brightness in the neighboring middle band, resulting in the perception of darker grey on the left side of the middle band. Similarly, the dark grey on the right side of the upper and lower bands exerts inhibitory effects on the darkness in the middle band, resulting in the perception of brighter grey on the right side of the middle band.

Such contrast enhancement strengthens the difference among observed objects, and presumably supports an animal to notice an object in a noisy or unclear environment, although there is a risk of possible misinterpretation or misunderstanding of the nature of the object as it obscures the real physical properties.

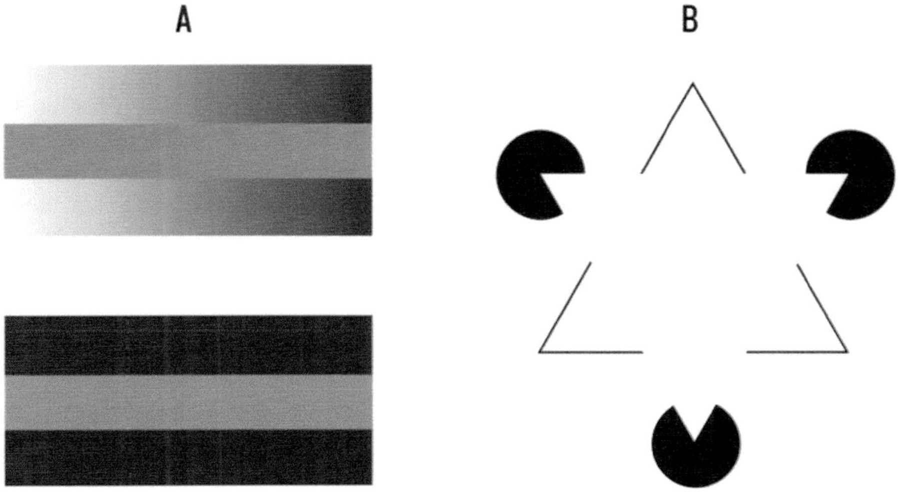

Figure 3. Contrast enhancement and filling-in. A, How do you find the darkness of grey in the middle band of a top figure? I guess that the left side looks darker than the right side. But, in reality the darkness of the middle band is uniform as shown in the bottom figure. This effect is caused by lateral inhibition. B, Do you recognize open and white triangles here? If so, they are caused by filling-in effects. This figure is called Kanizsa triangle (Kanizsa 1976) (from Hirano 2019, Purves et al. 2019).

Filling-in

Filling-in is a phenomenon in which the lacking parts of an object are filled automatically by the activities of neuronal circuits in the brain (Paradiso et al. 2006). An example of filling-in can be experienced in Figure 3B. Here, it appears that an open triangle surrounds three lines with one vertex at the top, and that an overlying white triangle has one vertex at the bottom. But actually, there is no real triangle in the figure. The matching of directions and positions of separated lines make us feel that the lines are connected, but concealed by another object, a white triangle in this example.

Neuronal mechanisms of filling-in have not been elucidated. Presumably we supplement the lacking information based on previous experiences, and this processing seems to take place automatically in the neuronal circuits of the brain. Filling-in supports recognition of objects with incomplete information, although there is a risk of misrecognition as information about unseen parts is lacking and speculated.

Illusion

Neuronal pathways designed for efficient information processing are subjected to a risk of creating illusions (Plodowski and Jackson 2001). The perception of brightness in the middle band of Figure 3A is caused by lateral inhibition, and the triangle recognition in Figure 3B is caused by filling-in. These can be regarded as examples of illusion. Recent studies reported that AI-assisted decoding of fMRI data obtained from the cerebral cortices of a research participant during the presentation of a visual image which could cause illusion, yielded reconstructed images having characteristics of illusion (Cheng et al. 2023).

The recognition of an object in ambiguous figures (Figure 4 of Chapter 5) might also be an illusion if the unrecognized object is real. Thus, illusion and the following misunderstanding cannot be avoided by our brains, which have a number of mechanisms to support and/or facilitate fast recognition of objects to speed up decision making and/or reaction. However, these processes sacrifice accurateness. Thus, we need to be careful about recognition in certain cases.

Information Transmissions with Line Connections and with Spreading Factors

Most information transmission between neurons occurs at synapses formed by closely apposed pre- and postsynaptic structures. Thus, information transmission takes place in a hard-wired one to one connection between a pair of neurons, which is somewhat similar to the working mechanism of computer systems. On the other hand, hormones have effects on various cells throughout the body simultaneously through the blood flow. There are also chemical substances which are secreted from certain cells and influence their surrounding cells. Such a way of information transmission is called paracrine transmission (Alberts 2023).

Amine neuromodulators such as dopamine, norepinephrine, etc. act as paracrine substances in the brain. In such cases, amines influence a large number of neurons simultaneously, resulting in the changes of activity levels of local neuronal circuits (Özçete et al. 2024). In sleep states which are regulated by neuromodulator molecules including amines, the neuronal activity patterns of the whole brain are altered from an awakened state, and functions of neuronal circuits are largely changed. Sleep states are explained further in chapter ten. Such a way of controlling elements in an information processing system has not been used in computer systems as far as I know. The functions of each neuromodulator molecule are explained in chapter nine.

Challenge for a Change by the Brain

Animals including humans sometimes choose an action whose results are uncertain as explained in chapter six. An expected value of rewards is usually low when an individual chooses an action whose outcome is uncertain. However, such a choice of behavior occasionally results in getting somewhat unexpected reward, and is useful to alter or adjust behavioral selection in a changing environment such as the case explained with Figure 2 of Chapter 6.

Computer systems are not designed to choose an output with a low expected value of rewards. AI can be designed to change or adjust its outputs after getting information of repetitive failures. However, most programmers of AI are unlikely to design to choose an uncertain and/or stochastically unfavorable outputs, when there is an output with a higher expected value of rewards or with a lower expected values of demerits or risks, as most users if not all, are unlikely to expect gambling reply from AI or a computer system.

Stability and continuity of performance is crucial in a computer system and also in animals. However, stability or unchanged responses might result in undesirable outcomes when some changes occur in the environment. Thus, stability and change in behavioral choices need to be balanced adequately. Animals seem to practice such balancing by occasional choice of behavior whose outcome is uncertain or a gamble, and the brain seems to have a system or machinery to support such a choice, whereas to appropriately design a computer system to give an output whose outcome is uncertain seems difficult or inappropriate, although to design AI which changes or adjusts outputs based on new data may be feasible.

The Significance of Diversity for The Survival of The Species

There are various people with different characteristics and they choose different behavioral actions in a similar situation. The diversity of individuals in an animal species seems to support the survival of the species as a whole in some cases, because some individuals making different choices of behavior may survive in certain conditions while others cannot make such choices. This situation might be somewhat analogous to the choice of behavior, the outcome of which is uncertain, as has been discussed in the previous section.

It has been reported that there are frogs who adopt different strategies for mating (Luo 2021). Most choose physical competition or fighting with male rivals in order to mate with female frogs, whereas there are a few frogs who hide while other frogs are fighting and wait for a good timing. And at a moment when other male frogs are concentrating on fighting or are exhausted, the male frog goes to a female and mates. Thus, such a frog can also have its offsprings.

Verbal communication and cooperation among individuals seems to have promoted the prosperity of human beings. Sympathy towards other people would have helped in cooperation. However, it has been reported that there are people called psychopaths and sociopaths (Glenn et al. 2013, De Brito et al. 2021). According to Cambridge Learner's Dictionary psychopathy is defined as "a condition in which someone has no feeling for others and does not feel bad about anything they have done", whereas a sociopath is "a person who is completely unable or unwilling to behave in a way that is acceptable to society". Both psychopathy and sociopathy are regarded antisocial personality disorders, although they are highly heterogeneous. There seem to be a large number of individuals in the border areas between normal and antisocial disorders. Apparently, these people might be thought troublesome. However, a person at the border of normalcy and psychopathy, in other words a normal person with psychopathic tendency, might be able to make adequate decision without being disturbed by strong emotions and might have more opportunity to survive in some risky conditions. Psychopathy is explained further in chapter eleven.

I wonder whether such diversity in strategies of action selection by individuals are recommended in computer systems or AI. If it happens, I doubt whether it can be appropriate. I feel that it is likely to be dangerous for the society. A limited number of people with light psychopathic or sociopathic tendency could sometimes contribute to the society, but if the number of such people is large and influential, society might become too unstable.

Singularity

According to Cambridge Learner's Dictionary, singularity is defined as "an event that some people believe will happen in the future when technology becomes much more intelligent than humans" or "the point in time when advances in AI can create machines that are smarter than humans". It is also said that "a physicist warned in 2014 that the development of full artificial intelligence could spell the end of the human race, but many of his peers don't expect a technological singularity any time soon." I wonder what is the definition of intelligence and smartness. In certain aspects of intelligence or information processing I think that AI has already had better ability than humans. For example, in some board games such as chess and go, AIs have already beaten human champions. I also think that the development of AI and AI agents is likely to reduce human jobs, including some kinds of management.

However, I would like to note again that humans are living beings having their own purpose, that is the survival of their selves and species, and that AI does not have its own purposes other than those set by humans at present. In imagination, it might be possible to believe that AI is designed to come to have its own purposes by machine learning using various information and data. But even in such a case I think that the goal or the final aim needs to be given. What should be set as a final goal or what should be given the first priority? Is it the merit of humans or AI? I wonder what can be the merit of AI without life. Merits of humans are also unclear or ambiguous. Merits of which populations should be given the highest priority, which of the short-term effects or long-term effects should be given higher priority, etc. are questions difficult to be answered. In most cases there are conflicts between merits and demerits, short-term and long-term effects, and/or among various populations, and there seem to be no correct answers or no solutions to which everyone can agree. These problems are discussed further in chapter twelve.

Summary and Conclusion

Overall, the design of a computer system and the brain are somewhat similar. Both of them are designed to process information that they get and/or is stored. Both have peripheral inputs and outputs devices, processing units and memories. There is a central information processing system which works to process input and stored information in both the brain and a computer system. The brain evaluates some pieces of information for decision making using the attention system, and the metacognitive processes or conscious thinking using the working memory. In a computer system CPU performs calculation and processes data in the cash memory. The working memory in a human and the cash memory in a computer are somewhat similar. Both the human brain and a computer system have different types of memories. The working memory and the cash memory are transient and are not retained, whereas some other memories are kept for a long time.

The materials composing the brain and a computer system are very different. The brain is mainly made from organic materials such as proteins, lipids, etc., whereas a computer system is made from inorganic materials such as semiconductor, etc. The differences in the materials of the processing elements result in differences in the speed of information processing between the brain and a computer. One step is about 0.1 nsec in a computer, whereas synaptic transmission needs around 1 msec. An individual brain has a limited lifespan, whereas a computer system does not have a certain lifespan. Death or the limited lifespan of an individual animal influences the purpose of its brain. The brain seems to have evolved so that animals behave in order to survive as individuals, and/or to contribute to a group to which it belongs. Emotion systems also seem to have developed to contribute for the similar purpose. On the other hand, all computer systems or AI have been designed to perform some specific purposes that some humans set, and do not have anything like emotional systems.

Faster information processing is advantageous for an animal to survive and to win in competition with rivals or enemies. Therefore, animals have developed several ways to speed up information processing in the brain at the expense of accuracy. Contrast enhancement with lateral inhibition and filling-in are examples. They contribute to fast recognition of an object in a noisy environment or with insufficient information, and support fast decision making and responses. However, they are accompanied with risks to have illusion and misunderstanding.

The brain also uses different ways of information processing such as information transmission using spreading factors in addition to hard-wired connections, and slower and longer-lasting processing of information through GPCRs and intracellular signaling cascades. The brain also chooses decisions with the expected outcome of which is uncertain. They could support fast change of behavior when the circumstance or environment have changed. Such a gambling choice of behavior and also diversities of characteristics of individuals in a species seem to contribute to the survival of the species as a whole. Neuronal circuits in the brain include complicated loop structures, and show large changes in functioning states depending on spreading neuromodulator molecules.

References

Alberts, B., K. Hopkin, A. Johnson, D. Morgan, K. Roberts, P. Walter et al. 2023. Essential Cell Biology. 6th Ed. Norton, New York, NY.

Cambridge Learner's Dictionary. 4th Ed. 2012 Cambridge Univ. Press, Cambridge, U.K.

Cheng, F.L., T. Horikawa, K. Majima, M. Tanaka, M. Abdelhack, S.C. Aoki et al. 2023. Reconstructing visual illusory experiences from human brain activity. Sci. Adv. 9: eadj3906.

De Brito, S.A., A.E. Forth, A.R. Baskin-Sommers, I.A. Brazil, E.R. Kimonis, D. Pardini et al. 2021. Psychopathy. Nat. Rev. Dis. Primers 7(1): 49.

Foley, J.M. 2019. Lateral effects in pattern vision. J. Vis. 19(9): 8.

Glenn, A.L., A.K. Johnson and A. Raine. 2013. Antisocial personality disorder: a current review. Curr. Psychiatry. Rep. 15(12): 427.

Hirano, T. 2013. Long-term depression and other synaptic plasticity in the cerebellum. Proc. Japan Acad. Series B 89: 183–195.

Hirano, T. 2019. What is the Brain for? Behavioral Selection and Neuroscience in the Era of AI (Japanese) Kyoto University Press, Kyoto, Japan.

Hirano, T. 2021. What is Myself? Neuroscientific Consideration of Self and Mind (Japanese) Gentousya Press, Tokyo.

Huguenard, J.R. and D.A. McCormick. 1992. Simulation of the currents involved in rhythmic oscillations in thalamic relay neurons. J. Neurophysiol. 68: 1373–1383.

Ito, M. 2001. Cerebellar long-term depression: characterization, signal transduction, and functional roles. Physiol. Rev. 81: 1143–1195.

Ito, M. 2006. Cerebellar circuitry as a neuronal machine. Prog. Neurobiol. 78: 272–303.
Kanizsa, G. 1976. Subjective contours. Sci. Am. 234: 48–52.
Luo, L. 2021. Principles of Neurobiology. 2nd Ed. CRC Press, Boca Raton, FL.
Masland, R.H. 2012. The neuronal organization of the retina. Neuron 76: 266–280.
Özçete, Ö.D., A. Banerjee and P.S. Kaeser. 2024. Mechanisms of neuromodulatory volume transmission. Mol. Psychiatry. doi: 10.1038/s41380-024-02608-3.
Paradiso, M.A., S. Blau, X. Huang, S.P. MacEvoy, A.F. Rossi and G. Shalev. 2006. Lightness, filling-in, and the fundamental role of context in visual perception. Prog. Brain Res. 155: 109–123.
Patterson, D.A. and J.L. Hennessy. 2012. Computer Organization and Design. 4th Ed. Elsevier, Amsterdam, Netherlands.
Plodowski, A. and S.R. Jackson. 2001. Vision: getting to grips with the Ebbinghaus illusion. Curr. Biol. 11: R304–306.
Purves, D., G.J. Augustine, D. Fitzpatrick, W.C. Hall, A.S. LaMantia, R.D. Mooney et al. 2019. Neuroscience. 6th Ed. Sinauer, Sunderland, MA.
Raman, I.M. and B.P. Bean. 1999. Ionic currents underlying spontaneous action potentials in isolated cerebellar Purkinje neurons. J. Neurosci. 19: 1663–1674.

8

Emotions and Feelings

What are Emotions and Feelings?

What are emotions? According to Cambridge Learner's Dictionary, emotion is defined as "strong feeling, such as of love, anger, fear, etc." Then what is included in the, etc.? Happiness, pleasure, expectation, surprise, sadness, hatred, disgust, grudge, anxiety, envy, jealousy, shame, admiration, pride and compassion may be included. There can be many other feelings too. There are a number of different definitions of emotions, and the complete definition is still a matter of debate (Damasio and Carvalho 2013, Malezieux et al. 2023). According to the above definition of the dictionary, emotions are nearly equal to feelings, although some scientists ascribe different meanings to emotions and feelings (Damasio and Carvalho 2013). According to an example of alternative definitions, feelings are primarily caused by the information coming from interoceptive systems and include hunger, thirst, pain, fear, etc., whereas emotions are caused by information coming from or based on external stimuli or recall, and include disgust, fear, anxiety, anger, sadness, joy, shame, etc. Fear belongs to both, as it could be caused both by the pain of a body part and by the existence of a predator, etc.

Both emotions and feelings are subjective conscious experiences for us, but they are also accompanied with some objective changes of the body states such as the heart beat rate, blood pressure, sweating, facial expression, etc. Some researchers define emotions as functional states that are adaptive for the organism, and differentiate these functional emotion states from the conscious percepts of emotions, which can be assessed only in humans (Malezieux et al. 2023). Such a definition supports experiments on emotions using animal species, in which emotions are assessed by behavioral responses and/or the biophysical parameters of an animal such as the heart beat rate.

An idea has been proposed that not only are emotions accompanied with body state changes, but that the latter also influences subjective conscious emotions (Critchley and Harrison 2013). An interesting report was published relating to this notion (Dutton and Aron 1974). A male research participant was asked to fill out questionnaires by an attractive female interviewer either on a fear-arousing suspension bridge or on a non-fear-arousing bridge. The interviewer also gave her telephone number to the participants and told them that she would inform them about

DOI: 10.1201/9781003518822-8

the results of experiment, if they were interested and they called her. More than eighty participants were tested. Few participants, who were on the non-fear-arousing bridge made the phone call, whereas the majority of participants who were on the suspension bridge called the interviewer. These results have been interpreted to mean that in a fearful condition the heart beat rate of a participant became higher, which made him feel that he was attracted to the interviewer.

Most emotions can be classified to positive ones such as pleasure, love and admiration, and negative ones such as fear, anger and sadness. There are also emotions which can be both positive and negative such as surprise, empathy, etc. In any case, emotions are self-centered experiences. In the following sections, most of emotions and feelings are explained.

Fear and Anxiety

Both fear and anxiety evoke defensive behavioral responses that work to avoid or reduce harm and support the survival of an animal. Fear is caused by a factual and acute threat, whereas anxiety is evoked by potential and/or anticipated threat (Davis et al. 2010, Tovote et al. 2015, Malezieux et al. 2023). The amygdala (Figure 1) has attracted much attention as the center of the brain responsible for fear. This is the main reason why neurons in the amygdala have been mainly studied in the auditory fear conditioning of rodents, as explained in chapter six. Monkeys, which had received lesions of the amygdala, showed uninhibited behavior in the presence of potentially dangerous objects. The neonatal lesion of the amygdala also affected the behavior in the adult monkey, such that the monkey readily took foods placed near a snake-like object, which elicited fearful responses in unoperated control monkeys (Bliss-Moreau et al. 2011).

There are also human cases. A patient S.M. with bilateral amygdala damage showed a deficit in experiencing fear triggered on observing other people's fearful faces (Feinstein et al. 2011). However, following studies showed that S.M. experienced fear when exposed to a high concentration of carbon dioxide (Feinstein et al. 2013). Such studies indicate that brain structures, other than the amygdala, are also involved in fear experiences. The ventral hippocampus, the medial prefrontal cortex and the bed nucleus of the stria terminalis (BNST) which is located ventral to the septum and anterior to the hypothalamus (Figure 1) (Lebow and Chen 2016), have been considered to play some roles in the evaluation of the presence or absence of threat. The periaqueductal gray (PAG) in the midbrain, the central amygdala, the hypothalamus, the lateral septum in the lower medial part of the forebrain, and the dorsal vagal complex in the medulla oblongata, are reported to be involved in the expression of fear responses (Figure 1) (Tovote et al. 2015, Malezieux et al. 2023). Most of the above brain areas, which are related to processing fear, are explained again within this chapter.

Compared with fear, anxiety is less well understood. Anxiety is evoked by potential and anticipated threats, and is characterized by sustained vigilance and apprehension. In a rodent anxiety is assessed by measuring the time during which an animal stays close to the walls avoiding the center in an open field, etc. The neuronal mechanisms to detect and process anxiogenic stimuli is more complex than those

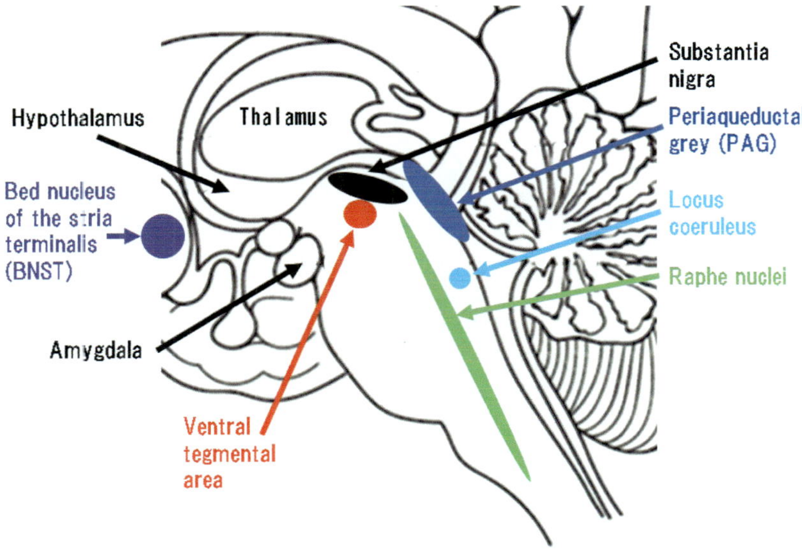

Figure 1. Some brain areas involved in processing emotions in the brainstem, the diencephalon, etc. are shown schematically in a sagittal projected plain. The substantia nigra and the ventral tegmental area containing dopaminergic neurons, the thalamus, the hypothalamus, the periaqueductal grey (PAG), the locus coeruleus containing noradrenergic neurons, the raphe nuclei containing serotonergic neurons, the amygdala and the bed nucleus of the stria terminalis (BNST) are shown (Lebow and Chen 2016, Venkatraman et al. 2017).

for fearful stimuli, although it has been considered that neuronal circuits underlying fear and anxiety overlap significantly (Figure 2) (Tovote et al. 2015, Malezieux et al. 2023).

The BNST receives inputs from the amygdala and is considered to be a part of the extended amygdala. It has been considered to play an important role in mediating anxiety (Tovote et al. 2015, Lebow and Chen 2016, Sherman et al. 2022). The optogenetic inhibition of the different areas of the BNST caused either anxiolysis or anxiety-like behavior. Ventral BNST sends both glutamatergic and GABAergic outputs to the ventral tegmental area (Figure 1). The activation of the former enhances anxiety, whereas the activation of the latter has anxiolytic effects (Tovote et al. 2015). Thus, neuronal elements which work in opposite directions are located close to one another in the BNST.

The involvement of the amygdala, the hippocampus and the medial prefrontal cortex in anxiety have also been reported. In addition, the septum has been considered to play a role in stress-induced anxiety. The involvement of other brain areas in anxiety has also been suggested (Figure 2). For example, it was reported that some neurons in the insular cortex show long-lasting activity increase during elevated anxiety. Such activity is required to keep a defensive state (Gehrlach et al. 2019). Some neurons in the ventromedial hypothalamus and the basolateral amygdala

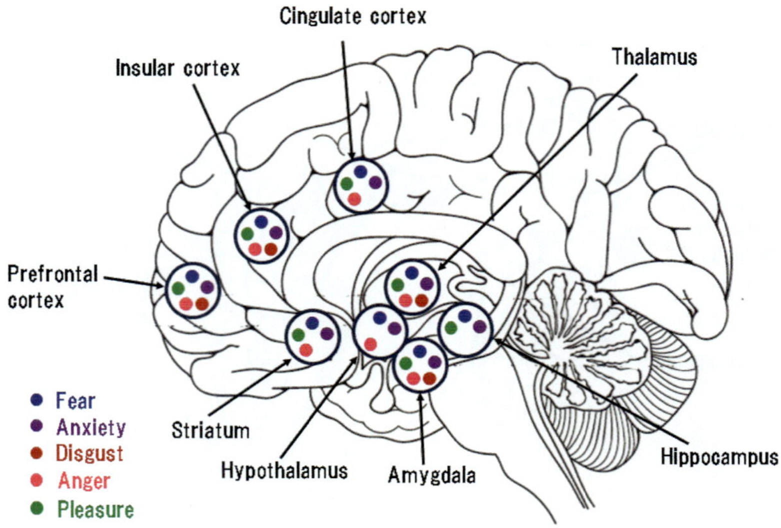

Figure 2. Brain areas implicated in multiple emotions such as fear, anxiety, disgust, anger and pleasure. They are the prefrontal, the insular and the cingulate cortices, the striatum, the thalamus, the hypothalamus, the amygdala and the hippocampus. Each area has subregions. The nucleus accumbens is a part of the ventral striatum. Other areas such as the ventral tegmental area, the bed nucleus of stria terminals (BNST), the periaqueductal gray (PAG), etc. (Figure 8-1) are also implicated in multiple emotions (Malezieux et al. 2023).

also show fear- and/or anxiety-related long-lasting activity increase (Malezieux et al. 2023).

The amygdala, the hippocampus, the medial prefrontal cortex and the BNST are reciprocally connected forming neuronal circuits (Tovote et al. 2015, Malezieux et al. 2023). The inclusion of the medial prefrontal cortex in this system explains emotional effects on attention, decision making and/or behavioral choice. The effects of stimulation or inhibition of the BNST or the amygdala on fear or anxiety-like behaviors in rodents indicate that such brain structures underly fear and anxiety-related behaviors. The above facts also suggest that the engram of the fear memory in the amygdala, identified using auditory fear conditioning, which is explained in chapter six, is only a part of a large system, which could however influence the function of the whole system. Strictly speaking, whether rodents experience subjective fear or anxiety when they show fear or anxiety-related behavior is unknown. The case study on the patient S.M. suggests that the amygdala underlies subjective experiences of fear in certain cases.

Anger and Aggression

Anger takes place when a goal of an individual is blocked, such as being interfered by other individuals, being treated unfairly or not receiving an expected reward. Anger might also occur due to social rejection, social isolation, frustration, etc. (Richard et al. 2022). Anger is not an avoidance-related but rather approach-related emotion, and anger sometimes elicits aggressive behaviors (Carver and Harmon-Jones 2009, Richard et al. 2022). Aggression is defined as hostile behaviors toward other animals, and seems to contribute to securing resources such as food, territory and/or a mating partner, and to defending oneself and the family. Thus, it can support competition with rivals and its survival (Lischinsky and Lin 2020). However, aggression costs a lot metabolically, and is potentially risky and can be fatal. Nevertheless, most animal species including humans show it occasionally.

Early experiments by Bard in 1928 reported that a cat in which the cerebral cortex was removed showed spontaneous aggression without any provocation (Richard et al. 2022, Malezieux et al. 2023). Then, Hess and Akert in 1955 reported that the direct electrical stimulation of a part of the hypothalamus elicited aggression-like behavior. Following studies have demonstrated that core neuronal regions processing aggression consist of the medial amygdala, BNST, the ventromedial hypothalamus and the ventral part of the premammillary nucleus in the hypothalamus, which are strongly interconnected. In addition, PAG plays an important role in the expression of aggressive behaviors (Figure 1). The medial prefrontal cortex and the lateral septum tonically suppress the hypothalamus to block aggression (Richard et al. 2022, Malezieux et al. 2023). It is known that human patients with lesions in their forebrain show enhanced aggression. However, the subpopulation of neurons in the medial prefrontal cortex promotes aggressive behaviors. It is to be noted that the neuronal circuits involved in aggression are situated near or even overlap with neuronal circuits mediating defensive behaviors elicited by fear (Figure 2) (Anderson 2016, Malezieux et al. 2023).

Aggression may not be accompanied by anger in some cases, and anger does not necessarily elicit aggressive behavior (Richard et al. 2022). Aggression-like behavior observed in Bard's experiment has been regarded as sham rage, because there was no target to attack. I wonder if a forebrain-lesioned cat experienced subjective anger when it showed sham rage.

Disgust and Avoidance

During evolution, the function of disgust has been considered to expand from the original role of defending the animal's body from toxins and diseases, to that of feelings towards various potential threats, including individuals violating social and/or moral norms (Chapman and Anderson 2012). The original form of disgust is thought to be distaste, which is caused by the ingestion of unpleasant-tasting substances which may be typically bitter, and results in spitting out of the substance. Materials and lives possibly contaminated with parasites that might cause infection and diseases such as feces, vomit, cockroaches, flies, etc., may also induce disgust, at least in humans. Most humans also experience disgust against people violating social norms and/or moral values such as those who lie, steal, harm others, etc. You may think that the above behaviors are more likely to cause anger rather than disgust, and that may be true in some cases. However, it is to be noted that aggressive behavior caused by anger costs a lot and is risky, whereas avoidance reaction caused by disgust is cost saving.

Disgust is accompanied with a particular facial expression such as the raising of the upper lip and the wrinkling of the nose in animals. Opening the mouth and lowering the brows also occur in some cases. These facial changes might work to reduce the amount of toxic material getting into the animal (Chapman and Anderson 2012). Upper lip raising is observed not only when expressing distaste, but also to express physical and moral disgust; suggesting their relationship (Chapman et al. 2009). The involvement of the anterior insular cortex and the amygdala in toxicity or disease-related disgust has been reported, and the anterior insular cortex is also implicated in the perception of facial expression of disgust (Figure 2). In a monkey the stimulation of the ventral anterior insular cortex causes spitting out or throwing away of preferred foods, as if the food has become distasteful (Caruana et al. 2010). It was also reported that optogenetic stimulation of the posterior insular cortex in a mouse can evoke facial expressions of disgust (Dolensek et al. 2020).

Although both disgust and fear are considered defensive states, their behavioral expressions and neural correlates are different (Malezieux et al. 2023). When experiencing fear, humans show faster eye movements, and an increase in nasal volume and in air velocity during breathing, whereas disgust causes opposite changes (Susskind et al. 2008). Thus, fear and disgust regulate the acquisition of sensory information in opposing directions. The involvement of the ventral pallidum in the basal ganglia and the nucleus accumbens, which is a front-ventral end of the striatum in the basal ganglia, in disgust, has also been proposed (Malezieux et al. 2023). The neuronal circuits of the basal ganglia are explained in a later section of this chapter.

Sadness and Regret

According to Cambridge Learner's Dictionary sadness is defined as "the feeling of being unhappy, especially because something bad has happened". Sadness has been considered to induce adaptive functions to cope with loss such as that of losing resources, status, friends, children, a romantic partner, etc. In humans, sadness is characterized by specific behavior, such as social withdrawal and lower reward seeking, and by typical facial expressions such as drooping eyelids, downcast eyes, slanting inner eyebrows, lowered lip corners, etc., and physiological bodily changes such as reduced heart beat rate, except during crying. Other animals also show somewhat similar facial expressions when they lose a member of the family or group. Thus, sadness seems to have developed during evolution. Sadness has been sometimes described as a psychological pain accompanied by additional feelings of loneliness, distress, depression, anxiety, grief and anguish, and could be a cause of major depressive disorders which are explained in chapter eleven. It has been suggested that mild sadness may afford considerable benefits including more accommodating, vigilant and externally-focused responses. The involvement of the cingulate and insular cortices, etc. in sadness have been reported (Arias et al. 2020).

According to Cambridge Learner's Dictionary regret is defined as "a feeling of sadness about something sad or wrong, or about a mistake that you have made, and a wish that it could have been different and better". Regret has significant influences on decision making including avoidance of decisions (Connolly and Zeelenberg

2002, Han et al. 2023), and it might contribute to improve future decision making and action selection in some cases.

Hate

Hate can be defined as intense dislike that might cause the elimination of others. It might involve dehumanization, denial of human qualities, and/or withholding empathy. It may be related to not only dislike but also anger (Mendez 2003). In one study research participants viewed the face of a person whom they hated and also faces of acquaintances for whom they had neutral feelings while fMRI was performed. The result showed the activation of the frontal cortex, the insular cortex and the putamen in the basal ganglia in the former case. By the way, the putamen and the insular cortex are also known to be implicated in romantic love (Zeki and Romaya 2008).

Pain

Pain works to protect the body of an animal from actual tissue injury. It might be better regarded as a sensation or a feeling rather than an emotion, as its signal is sent directly from peripheral terminals of the sensory axons in the body as explained in chapter four. Peripheral pain information is sent to the hypothalamus and the amygdala, and the latter conveys it to the insular and cingulate cortices (Damasio and Carvalho 2013, Gehrlach et al. 2019). However, pain can be felt without noxious sensory activations, and can be defined as an unpleasant subjective experience with a sensory and/or an emotional component. Pain coming from an amputated and non-existing limb has been reported, and called phantom pain (Collins et al. 2018).

Pain caused by the activation of a peripheral sensory system might be better called nociception (Baliki and Apkarian 2015, Gilam et al. 2020). It is also known that the activation of nociceptive pathways does not always elicit a subjective pain experience (Gilam et al. 2020). There is a top-down system to regulate the nociceptive pathway in the spinal cord as described in chapter four and in a later section of this chapter. The sensation of pain is processed in the brain, and the threshold and intensity of pain are strongly influenced by mood, attention, reward, etc. (Baliki and Apkarian 2015). Relief of pain can be a reward for animals and humans (Porreca and Navratilova 2017).

Pleasure and Reward

Pleasure is an internal reward for an animal. It occurs when an animal gets or experiences something good, for example consumption of food or mating, which contributes to its survival or the production of its offsprings, and promotes behaviors which are likely to result in such outcomes. Olds and Milner (1954) reported that the electrical stimulation of certain brain regions resulted in the reinforcement of behavior leading to the electrical stimulation (self-stimulation) in rats. This self-stimulation reward is very potent, so much so, that a hungry rat ignores foods, and continues to perform the behavior causing the self-stimulation (Olds 1956, Berridge and Kringelbach 2015). Following studies have shown that effective self-stimulation sites are those which induce the release of dopamine in the nucleus accumbens in the ventral basal ganglia (Hernandez et al. 2008). These findings suggest that dopamine is crucial in experiencing pleasure. However, this notion has been questioned by recent studies (Berridge and Kringelbach 2015, Malezieux et al. 2023).

Pleasure is a subjective experience as other emotions and qualia are. However, it is often accompanied with specific facial expressions in animals and humans. For example, sweet tastes cause relaxation of facial muscles and contended licking of the lips (hedonic reaction) in rodents, monkeys, apes and human infants. Such hedonic reactions were not diminished by dopamine blockade (Berridge and Kringelbach 2015). The microinjection of opioid into restricted hot spots in the nucleus accumbens enhanced the hedonic reaction to sweet taste in rats, whereas dopamine stimulation was without effect (Pecina and Berridge 2005, Berridge and Kringelbach 2015). The volume of hot spots is about 10% of the total volume of nucleus accumbens, and there are also cold spots to which the injection of opioid suppresses hedonic reactions. Similar hot spots have also been found in the ventral pallidum, the orbitofrontal cortex and the insular cortex, which are functionally connected. It has been reported that besides these neuronal structures, the anterior cingulate cortex and the amygdala are activated by rewards (Berridge and Kringelbach 2015, Malezieux et al. 2023). It has also been shown that Parkinson's disease patients with significant dopamine depletion show hedonic responses to sweet taste (Sienkiewicz-Jarosz et al. 2005). These findings cast doubts about the role of dopamine in hedonic experiences.

It has been proposed that there are separable processes regarding internal rewards. One is the motivation process for rewards called wanting, and another is the core reaction to hedonic stimulation or pleasure itself, called liking (Berridge and Kringelbach 2015, Hu 2016, Malezieux et al. 2023). Further, it has been considered that dopamine plays a role in wanting but not in liking. Thus, it has been suggested that dopamine strongly evokes a certain behavior towards getting rewards without experiencing pleasure. If so, I wonder why some drugs influencing the dopamine system such as cocaine and methamphetamine, could make people feel pleasant. Possible mechanisms have been discussed. A psychological explanation is that the elevated incentive salience induced by dopamine may be mistaken for pleasure itself. The enhancement of the incentive salience induced by a drug, might make other people, actions or events in the world all seem more attractive, and they might carry an aura of euphoria, even if one does not have a true hedonic experience. A neuronal explanation for why (for example) cocaine is pleasant, is that cocaine stimulates the secondary recruitment of endogenous opioid beyond raising dopamine release. It has been reported that dopamine-stimulating drugs recruit elevation of endogenous opioid in the nucleus accumbens (Berridge and Kringelbach 2015).

Dopamine has also been shown to be involved in reinforcement learning. It was proposed that the amount of dopamine release is correlated with a prediction error, that is the reward outcome minus the expected reward, and the dopamine signal conveys the feedback information which influences synaptic plasticity (Schultz 1998, Berridge and Kringelbach 2015, Hu 2016). Besides the neuronal structures mentioned above, the involvement of the lateral hypothalamus, the dorsal striatum, the ventral tegmental area, etc. in the pleasure neuronal system has been suggested (Figures 1 and 2) (Berridge and Kringelbach 2015). Dopamine is further explained in chapter nine.

Love and Attachment

Romantic and maternal love strongly influence behavioral choices, and contribute to maintaining the species (Zeki 2007, Feldman 2017). fMRI studies, etc. showed involvement of the medial insular cortex, the anterior cingulate cortex, the hippocampus, the amygdala and the nucleus accumbens in human attachment (Figures 1 and 2). In addition, neuromodulators, dopamine, oxytocin and vasopressin, have attracted attentions as molecules implicated in attachment in humans and animals. Oxytocin and vasopressin are actually peptide hormones, which are produced in neurons in the hypothalamus and are secreted from their axon terminals in the posterior lobe of pituitary organ (Figure 2B of Chapter 4) to blood vessels. Some neurons in the brain have receptors for oxytocin and vasopressin, and these hormones also work as neuromodulators, which are explained in chapter nine.

The cross-talk of dopamine and oxytocin systems in the nucleus accumbens has been considered to play a critical role in attachment toward a particular object such as a partner or a baby (Feldman 2017). In a monogamous prairie vole, oxytocin receptors in the nucleus accumbens have been implicated in the regulation of pair bond formation. A female prairie vole overexpressing oxytocin receptor in the nucleus accumbens showed enhanced formation of partner preference after cohabitation with a male (Ross et al. 2009).

Empathy, Sympathy and Compassion

As explained in chapter two, some animals show altruistic behaviors. A monkey helps another monkey so that the latter can obtain a food (Burkart et al. 2014). Rats avoid actions that harm other rats, and prior experience with pain enhances avoidance (Hernandez-Lallement et al. 2020). The pharmacological deactivation of the anterior cingulate cortex reduced the harm aversion. Rats might have a kind of theory of mind, empathy, sympathy and/or compassion. Empathy is thought to rely on the prefrontal, the cingulate, the insular cortices and the amygdala (deWaal and Preston 2017, Paradiso et al. 2021).

By the way, empathy, sympathy and compassion are related but different concepts (Lamm et al. 2019). According to Cambridge Learner's Dictionary, empathy is defined as "the ability to share someone else's feelings or experiences by imagining what it would be like to be in that person's situation", sympathy as "understanding and care for someone else's suffering", and compassion as "a strong feeling of sympathy and sadness for the suffering or bad luck of others and a wish to help them". Thus, empathy is directed to another individual who is either in a positive or a negative situation, whereas sympathy is directed to an individual who is in a negative situation. Compassion is accompanied with a wish to help a suffering individual.

Envy and Schadenfreude

Humans compare self and others in ability, social status, etc., and sometimes experience envy when the other person is superior, and feel schadenfreude (pleasure derived from another person's misfortune) occasionally when an envied individual fails or experiences an unfortunate affair. In one research on human envy and schadenfreude, research participants read a scenario in which the protagonist and three other persons appeared. They were asked to imagine that the protagonist was themselves (Takahashi et al. 2009). In the scenario for a male participant, the protagonist was a male student with an average ability, quality, and social status. The three other characters in the scenario, were students A, B and C. A was a male student, had superior quality, ability and was highly relevant to the protagonist. A female student B also possessed superior quality and ability, but low relevance. A female student C had average ability, quality and low relevance. Participants reported the strongest envy for A followed by B and C, which was correlated with the strength of the neuronal activity in the anterior cingulate cortex by fMRI. In the next scenario, a misfortune happened to student A and student C. Participants reported stronger pleasure at the misfortune that occurred to student A, which was accompanied with the activation of the ventral striatum in the basal ganglia. Thus, envy seems to be stronger toward an individual possessing superior attributes as compared to oneself and higher relevance.

Envy or jealousy does not seem to be specific to humans. At least dogs show behaviors presumably reflecting jealousy (Harris and Prouvost 2014). Envy is not always negative and malicious, and there is also positive and benign envy toward the target individual which can promote job engagement in the workplace (Battle and Diab 2022). Envy might have evolved to strengthen the social status of an individual among competitors.

Startle and Surprise

An unexpected and sudden intense tactile, acoustic or visual stimulus elicits a startle, which is a kind of transient fear, and is accompanied with fast protective responses such as eyelid-closure, contraction of facial, neck and skeletal muscles, acceleration of the heart rate, and the arrest of ongoing behaviors (Koch 1999). We also experience surprise when we meet an unexpected event. Surprise is short-lasting, and strengthens emotions such as pleasure and sadness (Mellers et al. 2013). In humans a kind of event-related potential called mismatch negativity has been known to be a brain response to an unexpected event (Khouri and Nelken 2015).

Neuronal Circuits for Emotions

Subcortical regions such as the hypothalamus, the amygdala, the nucleus accumbens, and the cortical regions, such as the insular, the anterior cingulate and the medial prefrontal cortices, are involved in multiple emotions (Figure 2). In addition, the brain stem region PAG (Figure 1) is also implicated in multiple emotions (Malezieux et al. 2023). Panksepp (2011) proposed that emotions are processed at three levels. First, the primitive features of emotion are processed in the deep subcortical structures such as the brainstem, the basal ganglia, the thalamus, the hypothalamus and the amygdala, which are conserved in many animal species. This notion is supported by reports that emotion expressions are exhibited in humans and animals in the total absence of neocortex, and that activity manipulations of some hypothalamic neurons elicit emotion expressions. In the secondary processes, brain neuronal circuits comprising the hippocampus, the hypothalamus, the thalamus and the amygdala support the evaluation and integration of multifaceted stimuli, and also learning. At the third level, cognition and subjective appraisals of feelings are processed by the neocortical areas. However, emotions are not likely to be processed along a one-way stream, and there are many feedforward and feedback loops connecting different processing levels (Malezieux et al. 2023). Pessoa (2007) proposed that emotions are controlled by large-scale network interactions spanning large brain regions.

There are strong interactions between emotions and interoceptions, and the vagus nerves play important roles in the interactions. Interoceptive information, such as about the heart rate and breathing, is transmitted to the insular cortex through the vagus nerves, and influences emotions such as fear and anxiety. On the other hand, emotions affect functions of various internal organs such as the gastrointestinal tract, the heart, the lung, the immune system and the vascular system (Malezieux et al. 2023).

As described above, neurons working in opposite directions are found in neighboring regions in the neuronal structures involved in emotions; such as the BNST (Figure 1) processing anxiety, the medial frontal cortex controlling aggressive behaviors and the nucleus accumbens implicated in hedonic responses as described above. These neuronal organizations together with the overlap of neuronal regions involved in the processing of multiple emotions, might cause unexpected interactions among emotions (Figure 2).

I have been wondering why there are many people who seek to have fearful experiences such as riding a roller coaster, and why many people like to eat burning or spicy foods. Chili pepper contains capsaicin that activates nociceptive receptor TRP ion channel, which conveys a sensation of pain. Thus, some people seek fearful and/or painful stimulations. Nevertheless, they seem to experience a kind of pleasure from such behaviors. There are several neuronal structures processing not only fear and pain but also pleasure such as the amygdala and the insular cortex (Figure 2) (Malezieux et al. 2023). There might be some interaction between neurons processing different or opposite emotions in such areas. In the following sections I add explanations about major brain regions involved in the control of emotions.

Limbic Circuit

The term 'limbic" has a long history. It was first introduced by Willis in 1664 to designate a cortical border encircling the brainstem. In 1937 Papez proposed that emotion arises either from cognition entering the limbic neuronal circuit from the cerebral cortices through the hippocampus, or from visceral and somatic perceptions entering the circuit through the hypothalamus. Papez considered that the cingulate cortex might be the target area receiving information coming from the hypothalamus or the hippocampus for experiencing emotions. In 1948 Yakovlev proposed that the orbitofrontal cortex, the insular cortex, the amygdala and the anterior temporal lobe form a network underlying emotions and motivation (Figure 2 of Chapter 4). After that MacLean concluded that the limbic cortices, together with the limbic subcortical structures, are a functionally integrated system involved especially in emotions. However, more recently it has been proposed that there are separate systems in the brain for emotions and for episodic memory, each involving limbic and other structures (Catani et al. 2013, Rolls 2015). In following sections, several brain regions implicated in emotions in and outside of the limbic circuit are explained.

Insular Cortex

The insular cortex which is located inside of the cerebral hemisphere along the Sylvian fissure (Figure 2 of Chapter 4), has long been recognized as a brain center in various functions, such as sensory processing, bodily homeostasis, emotions, motivation and decision-making (Uddin et al. 2017, Zhang et al. 2024). The insular cortex has been divided along the anterior-posterior axis, yielding the anterior, medial, and posterior insular cortices. The anterior insular cortex is primarily involved in emotional processes and cognitive control such as decision making and interoceptive awareness. The posterior insular cortex is located close to the primary sensory regions such as the somatosensory, visual and auditory cortices, and involved in sensory processing including taste and pain perception, in which the insular cortex plays primary roles as explained in Chapter 4. The medial insular cortex exhibits mixed characteristics and possibly integrates their functions. Insular cortices can be also divided dorsal-ventrally into granular, dysgranular and agranular regions based on the relative presence or absence of the granule cell layer (layer IV, Figure 8 of Chapter 4). The agranular regions are considered to contribute to higher-order cognitive and emotional processing, whereas the granular regions are considered to be associated with sensory and perceptual processing. The dysgranular regions have been regarded as transitional areas supporting the integration of the diverse functions of the insular cortices. The agranular and dysgranular cortical regions are also found in other cerebral cortical regions, including the prefrontal cortices. The insular cortex receives inputs from sensory and motor cortices, the amygdala, etc. and sends outputs to various cerebral cortical regions, the striatum, the thalamus, the amygdala, etc. There are also rich intra-insular cortical connections (Zhang et al. 2024).

The insular cortex has been considered to work as a bidirectional bridge between sensation and emotions. In vivo calcium imaging studies have reported that neurons in the insular cortex encode various types of emotions including pleasure, disgust, fear and pain (Zhang et al. 2024). The involvement of the insular cortex in other emotions such as anxiety, disgust, anger, sadness and empathy has also been reported (Figure 2) (Malezieux et al. 2023, Zhang et al. 2024). It has been proposed that the insular cortex contributes to body awareness which is defined as consciousness of the self in physical space and also to time perception which is enabled through sensing dynamically changing body processes. The insular cortex is considered to process the temporal information originating from alterations in bodily states and integrates information from the entire body continuously, and evaluates the passage of time based on the internal physiological rhythms such as heart rate, breath rate, etc. The insular cortex also seems to be involved in risk evaluation and decision making (Zhang et al. 2024).

Cingulate Cortex

The cingulate cortex is located inside the cerebral hemisphere overlying the corpus callosum (Figure 2B of Chapter 4), and is divided to the anterior, mid- and posterior cingulate cortices with subdivisions (Vogt 2005, Rolls 2019). The anterior cingulate cortex is considered to receive information from the orbitofrontal cortex about reward and non-reward outcomes, and to connect rewards to actions. The posterior cingulate cortex receives spatial and action-related information from parietal cortical areas and send outputs to the hippocampal systems, which might allow to perform action–outcome learning (Rolls 2019).

fMRI study reported that a subregion of the anterior cingulate cortex is activated during sad events, whereas another subregion is activated when one experiences a happy emotion. Pain responses are also detected in the subregions of the anterior cingulate cortex. Some regions in the mid-cingulate cortex respond to noxious stimuli. The mid-cingulate cortex receives inputs from the amygdala and its activities seem to be associated with fear. There are overlaps between the fear and pain sites in the cingulate cortex, implying involvement of these regions in avoidance behaviors. The posterior cingulate cortex is suggested to be involved in the visuospatial orientation, its subregions might be involved in orienting the body toward innocuous and noxious somatosensory stimuli. Some regions of the posterior cingulate cortex show high neuronal activities during happiness (Vogt 2005). The involvement of the cingulate cortex in anxiety and anger has also been suggested (Figure 2) (Malezieux et al. 2023).

Prefrontal Cortices

The prefrontal cortices are the anterior areas of the frontal lobe of the cerebrum, and are implicated in emotions, perception, motivation, attention, working memory and decision making (Figure A of Chapter 4) (Carlén 2017). The involvement of the medial prefrontal cortex and the orbitofrontal cortex in processing emotions has been reported.

A lesion in the orbitofrontal cortex, which is the ventral part of the frontal cortex and located above eye orbits, causes increased aggression and reduces fear-related behaviors in animals (Rudebeck and Rich 2018). The orbitofrontal cortex receives information from all sensory modalities, sends outputs to the nucleus accumbens, etc., and has bidirectional neuronal connections with other prefrontal cortices, the cingulate cortex, the insular cortex, the amygdala, the thalamus, the hypothalamus, etc. (Wallis 2007, Rudebeck and Rich 2018). These neuronal connections together with reports about effects of lesions of the orbitofrontal cortex, suggest its involvement in processing and/or control of emotions such as pleasure, fear, anxiety, disgust, anger, etc. (Figure 2) (Malezieux et al. 2023).

The damage to the human orbitofrontal cortex does not impair cognitive abilities but impairs everyday decision making. It was reported that a patient whose bilateral orbitofrontal cortices had been damaged by an operation to remove the tumor, retained his intelligence, memory, reading and writing comprehension, verbal fluency, visuospatial abilities, facial recognition, etc., but that he began to have difficulty in making simple decisions. For example, deciding where to dine took hours as he agonized over every possible consideration such as the menu, the seating arrangement and the atmosphere. He also showed difficulty in changing behavioral choices after having negative outcomes from certain behaviors. (Wallis 2007). On the other hand, other patients with orbitofrontal cortical damages showed reduction of emotional and behavioral inhibition. Based on these reports it was suggested that the orbitofrontal cortex exerts inhibitory control on behaviors. However, this hypothesis has been questioned, and a newer view about functions of the orbitofrontal cortex is that it encodes or stores specific stimulus–reward associations (Rudebeck and Rich 2018).

Wallis (2007) proposed that the orbitofrontal cortex plays a key role in processing reward information through integrating multiple sources of information regarding the reward outcome to derive a value signal. It is then held in the working memory and used by the lateral prefrontal cortex to plan and organize behaviors toward obtaining the outcome, and by the medial prefrontal cortex to evaluate the overall actions in terms of their success and required efforts.

Amygdala

The amygdala is associated with emotions and motivation, and involved in processing both fearful and rewarding stimuli (Janak and Tye 2015). Developmentally the amygdala is formed from the telencephalon, and found inside the temporal lobe of the cerebral cortices (Figure 1, also Figure 2B of Chapter 4). The amygdala consists of several subdivisions; the basolateral amygdala made up of the lateral, basal and basomedial nuclei, and the central amygdala made up of a lateral subdivision and a medial subdivision. The basolateral amygdala consists of glutamatergic principal neurons and inhibitory interneurons. Neurons in the central amygdala are primarily GABAergic. Neurons in the lateral subdivision project to the medial region. Neurons in the lateral amygdala receive sensory information from the thalamus and sensory cortices, and send outputs within the basolateral amygdala and to the central amygdala. The basolateral amygdala is reciprocally connected with the prefrontal cortices, the sensory cortices and the hippocampus. The basolateral amygdala sends unidirectional outputs to the striatum in the basal ganglia, especially to the nucleus accumbens and also to the BNST (Figure 1) (Janak and Tye 2015). The central amygdala influences the activity of the hypothalamic–pituitary–adrenal axis as explained in chapter four (Lebow and Chen 2016).

The amygdala is a rapid detector of aversive environmental stimuli and situations, and has been considered to produce affective behavioral states causing adaptive responses to potential threats. The involvement of the lateral and central amygdala in auditory fear conditioning has been reported, and explained in chapter six. In addition, the involvement of the amygdala in reward processing has also been reported. A recording of neuronal activities in the basolateral amygdala found populations of valence-selective neurons, which are excited by a fear cue and do not respond to a reward cue, or show inhibition due to the reward cue, and vice versa. I note that there are also neurons responding to both fear and reward cues (Janak and Tye 2015). The amygdala has also been implicated in other emotions such as anxiety, anger and disgust (Figure 2) (LeDoux 2007, Malezieux et al. 2023).

The Bed Nucleus of the Stria Terminalis (BNST)

The bed nucleus of the stria terminalis (BNST) is sometimes referred to as the extended amygdala (Figure 1). It consists of more than ten sub-nuclei, and is located ventral to the septum, anterior to the hypothalamus and is enclosed on both sides by the pallidum or the globus pallidus of the basal ganglia. The BNST is sexually dimorphic, and human males have BNST that is 2.5 times greater in volume than females. BNST is involved in stress responses, sustained fear or anxiety, attachment, aggression, etc. Mood and arousal states seem to be processed via BNST's connections to the amygdala, the raphe nuclei, the ventral tegmental area (Figure 1) and the medial prefrontal cortex. Its connections with the hypothalamus seem to allow monitoring of feeding, drinking and fluid maintenance, and those with the lateral septum and the medial amygdala seem to allow coordination of reproductive behaviors as part of the social behavioral network. In the BNST some neurons have a corticotrophin-releasing factor, and influence the activities of the hypothalamic–pituitary–adrenal axis. The BNST is implicated in some psychiatric disorders such as anxiety disorders, post-traumatic disorders, psychopathy, etc. which are explained further in chapter eleven (Lebow and Chen 2016).

Basal Ganglia

The basal ganglia are a group of subcortical nuclei involved in motor control, cognition and emotions (Figure 3, also Figure 2 of Chapter 4). The striatum composed of the caudate nucleus and the putamen, and the subthalamic nucleus are input nuclei (Figure 3). The major inputs to the basal ganglia derive from the cerebral cortices, the midbrain and the thalamus. The internal segment of the globus pallidus or the pallidum, and substantia nigra pars reticulata are the output nuclei. It is to be noted that actually the subthalamic nucleus belongs to the subthalamus and the substantia nigra (Figure 1) belongs to the midbrain, although they are functionally highly related to the basal ganglia. The input and output nuclei are interconnected by the direct and indirect pathways (Figure 3B). The cerebral cortex, the basal ganglia and the thalamus communicate with each other forming loop structures. The nucleus accumbens which plays a critical role in processing emotions, is the ventral part of the striatum. About 90% of striatal neurons are GABAergic medium spiny projection neurons (Afifi 2003).

Inputs to the basal ganglia from the cerebral cortices and the thalamus are glutamatergic. On the other hand, the striatum receives strong dopaminergic inputs from the substantia nigra pars compacta. Actually, the concentration of dopamine is the highest in the striatum within the brain. The dopaminergic projection from

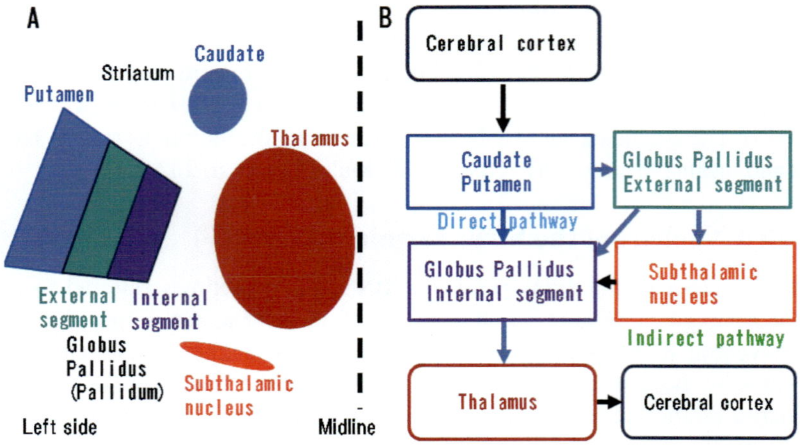

Figure 3. The basal ganglia. A, Main structures of the left basal ganglia including striatum consisting of the caudate and the putamen, the globus pallidus or the pallidum consisting of the external and the internal segments, are schematically presented together with the subthalamic nucleus and the thalamus in a cross section. The caudate and the putamen are actually connected outside of the plane. These structures are found inside of the cerebral cortices (Figure 2 of Chapter 4). B, The direct and the indirect pathways of the basal ganglia. GABAergic inhibitory neurons in the striatum (the caudate and the putamen) receive major inputs from the cerebral cortices, the midbrain and the thalamus, and send outputs to either the internal segment of the globus pallidus (direct pathway), or the external segment of the globus pallidus (indirect pathway), which in turn sends inhibitory outputs to the internal segment of the globus pallidus directly or indirectly through the subthalamic nucleus. Black and blue arrows indicate excitatory and inhibitory synaptic transmission, respectively. GABAergic neurons in the internal segment are intrinsically active and send outputs to the ventral anterior and the ventral lateral nuclei of the thalamus, which send information to the cerebral cortices. The above direct pathway in the basal ganglia consists of transiently active inhibitory neurons followed by tonically active inhibitory neurons, and works as a disinhibitory pathway. On the other hand, in the indirect pathway there are three serially connected inhibitory neurons, and the net output of the pathway is inhibitory. Neurons in the subthalamic nuclei are excitatory. The substantia nigra pars reticulata also work as an output neuronal region as the globus pallidus internal segment. The subthalamic nucleus also receives excitatory inputs from the cerebral cortices. The substantia nigra and the neuronal pathway from the cortex to the subthalamic nucleus are not shown in this figure for simplicity (Purves et al. 2019).

the substantia nigra pars compacta to the dorsal striatum has a net excitatory effect on striatal neurons that send GABAergic outputs to the internal segment of the globus pallidus and the substantia nigra pars reticulata in the direct pathway (Figure 3B), and the dopamine projection has a net inhibitory effect on the striatal neurons that send GABAergic outputs to the external segment of the globus pallidus in the indirect pathway. GABAergic neurons in the external segment of globus pallidus project to the internal segment of globus pallidus directly and indirectly through the subthalamic nucleus (Figure 3B). The glutamatergic subthalamic projections to the output nuclei of the basal ganglia, which are the internal segment of globus pallidus and substantia nigra pars reticulata, interact with many neurons, whereas striatal GABAergic projections are focused on a single neuron. Thus, the subthalamic nucleus in the indirect pathway provides divergent excitation, and the striatum direct pathway provides focused inhibition in the output nuclei. Therefore, focused inhibition and surround excitation of neurons in the output nuclei of basal ganglia take place (Afifi 2003, Purves et al. 2019).

The major targets of output neurons in the basal ganglia are some nuclei in the thalamus, and they are inhibited by the GABAergic outputs (Figure 3B). The activation of the direct pathway seems to lead to a net disinhibitory effect on the thalamus and to enhance motor actions, whereas the activation of the indirect pathway seems to lead to increased inhibition in the thalamus, resulting in decreased motor activity. The targeted thalamic nuclei provide the link between the basal ganglia and the cerebral cortices including the primary and higher order motor cortices, the prefrontal cortices, the cingulate cortex, etc. (Afifi 2003).

The nucleus accumbens is a part of the ventral striatum and is largely involved in reward and pain processes. The nucleus accumbens receives glutamatergic inputs from the amygdala, the hippocampus, the thalamus and the prefrontal cortex, and dopaminergic inputs from the ventral tegmental area (Figure 1). The nucleus accumbens sends outputs to the ventral pallidum in the basal ganglia, the hypothalamus, the midbrain and some motor effector regions. The nucleus accumbens has been subdivided to a medial "shell" subregion and a more lateral "core" component (Floresco 2015, Yan et al. 2023). It was proposed that the nucleus accumbens supports obtaining motivationally relevant goals by promoting likelihood, efficiency and vigor of behaviors intended to obtain the goals, which can be rewards or avoidance of aversive consequences. The core subregion has been suggested to play a more prominent role in the promotion of an approach behavior toward motivationally relevant stimuli, while the shell subregion has been suggested to suppress irrelevant or nonrewarded actions so that goals may be obtained more efficiently (Floresco 2015). The involvement of the nucleus accumbens in fear, anxiety and anger has also been suggested (Figure 1) (Malezieux et al. 2023).

Thalamus

The thalamus is a bilateral midline diencephalon structure consisting of more than forty nuclei (Figure 1 and Figure 2B of Chapter 4), and works as an integrative center. The thalamus has been recognized as an essential sensory relay station that filters information and sends outputs to the primary sensory cortices as explained in Chapter 4. Nevertheless, roles of the thalamus are complex and it also serves higher brain functions including processing emotions such as fear, anxiety, disgust, anger and pleasure (Figure 2), arousal, awareness, etc. (Georgescu et al. 2020, Malezieux et al. 2023).

Among the thalamic nuclei, the mediodorsal nuclei have bidirectional connections with the prefrontal cortices, and are involved in cognitive processes including, the working memory, attention control, planning, decision making, reward evaluation, emotions, sensory discriminations and learning. The mediodorsal nuclei also receive inputs from the amygdala, the internal segment of the globus pallidus in the basal ganglia, the locus coeruleus and the raphe nuclei. They also receive olfactory and pain information. They receive non-dopaminergic inputs from the ventral tegmental area and the substantia nigra pars reticulata (Figure 1). The mediodorsal nucleus sends outputs to the basal ganglia. Thus, there are bidirectional connections between the mediodorsal thalamic nuclei and the basal ganglia. It was also proposed that the mediodorsal thalamic nuclei receive inputs from the cerebellar nucleus and form the cerebello-thalamo-cortical pathway. I note that some nuclei in the thalamus other than the mediodorsal nuclei are also implicated in processing emotions (Georgescu et al. 2020).

Neurons in the mediodorsal thalamic nuclei show tonic and bursting modes of action potential firings, and seem to affect the level of consciousness. When the membrane potential of a mediodorsal nuclear neuron is above –55 mV, the application of brief depolarizing current induces repetitive fast firing of action potentials, whereas if the membrane potential is less than –60 mV, the same stimulus triggers a burst of action potentials, which are followed by a 150–180 msec refractory period.

In sleep, the inputs from the reticular nucleus in the pons hyperpolarize the thalamic neurons and cause bursting, which seems to result in the generation of high-amplitude slow waves in the electroencephalogram, that is explained in chapter ten. When thalamic neurons are in a bursting mode, information transmission is disrupted. In contrast, when the thalamic neurons are in a tonic mode, sensory information can be transmitted to the cerebral cortices. During tonic firing, these neurons can fire synchronously at about 40 Hz which corresponds to γ frequencies, that are explained in chapters five and ten (Sherman 2001, Georgescu et al. 2020).

Hypothalamus

The hypothalamus located under the thalamus (Figure 1, also Figure 2B of Chapter 4) has regulatory circuits that support basic life functions and ensure the survival of an individual animal. It receives, processes and integrates sensory inputs to drive behavioral, autonomic and endocrine response pathways. The hypothalamus includes many nuclei and a large variety of neuronal types with different expression patterns of neurotransmitters, neuropeptides and transcription factors, etc. The hypothalamus has neuronal connections with the thalamus, the amygdala, the hippocampus, the cerebral cortices and the brainstem (Fong et al. 2023).

The hypothalamus connects to the anterior and posterior pituitary organs via the pituitary stalk (Figure 2B of Chapter 4). Some neurons terminating in the posterior pituitary organs secrete arginine-vasopressin or oxytocin into the peripheral circulation. Other neurons secrete hypothalamic-releasing hormones, such as the corticotropin releasing hormone, the gonadotropin releasing hormone, etc., which go to the anterior pituitary through blood vessels and stimulate the synthesis and secretion of pituitary hormones such as the adrenocorticotropic hormone, the luteinizing hormone, the follicle-stimulating hormone etc. Thus, the hypothalamus is involved in the control of reproduction with the oxytocin and gonadotropin releasing hormone; in stress responses with the corticotropin releasing hormone through the hypothalamus-pituitary-adrenal axis; and in control of osmoregulation through vasopressin (Fong et al. 2023). The implication of oxytocin in attachment is explained further in chapter nine. In addition, the involvement of the hypothalamus in the expression of other emotions such as fear, anxiety and anger has also been suggested (Figure 2) (Malezieux et al. 2023).

The hypothalamus is also involved in the regulation of energy balance, sleep-wake cycles, body temperature, blood pressure, etc. The arcuate nucleus in the ventromedial hypothalamus plays a critical role in the regulation of the amount of food intake (see Chapter 4), as well as in other homeostatic roles of the hypothalamus. The suprachiasmatic nucleus in the hypothalamus plays a central role in generation of circadian rhythm which is explained in chapter ten.

Midbrain

The periaqueductal gray (PAG), the ventral tegmental area and the substantia nigra belong to the midbrain (Figure 1), although the substantia nigra is often mentioned as a part of the basal ganglia because of its strong connections to structures of the basal ganglia. The rostral part of the serotonergic raphe nuclei (Figure 1) and the cholinergic laterodorsal tegmental nuclei also belong to the midbrain (Venkatraman et al. 2017).

PAG plays a role in integrated behavioral responses to internal or external stressors. It is involved in the control of behavioral and autonomic responses to pain, fear and anxiety. PAG receives inputs from the prefrontal cortex, the cingulate cortex, the insular cortex, the amygdala, the hypothalamus, and the nociceptive pathways, such as afferents from the spinal dorsal horn and from the trigeminal nuclei in the brainstem. PAG also receives norepinephrine and epinephrine inputs. PAG sends outputs to the thalamus, the hypothalamus, the medulla oblongata and the spinal cord. Via its connections with different brainstem nuclei, PAG coordinates cardiovascular, respiratory, motor and pain modulatory responses. PAG stimulation strongly inhibits activities of some dorsal horn neurons in the spinal cord that relay nociceptive information, contributing to top-down suppression of pain sensation which is described in chapter four. PAG is also involved in vocalization, micturition, and thermo-regulation, and contributes to the control of arousal and REM sleep, which is explained in Chapter 10 (Benarroch 2012).

The ventral tegmental area plays critical roles in the control of reward-related or aversion-related behaviors. It receives inputs from various brain regions including the prefrontal cortex, the basal ganglia, the hypothalamus, PAG, BNST, the lateral habenula in the epithalamus and the raphe nuclei, and sends outputs to the nucleus accumbens, the frontal cortex and the lateral habenula. The primary projection neurons of the ventral tegmental area are dopaminergic, although there are glutamatergic and GABAergic projection neurons also. Some neurons there express and co-release multiple neurotransmitters such as dopamine and glutamate, dopamine and GABA, or glutamate and GABA. The outputs of the ventral tegmental area to the lateral habenula are not dopaminergic. Some putative dopaminergic neurons in the ventral tegmental area are activated by unpredicted rewards, and shift their activation to cues that predict reward following learning, and transiently decrease their action potential firing when an expected reward is not obtained. Some dopaminergic neurons are also excited or inhibited by aversive stimuli or by cues that predict an aversive outcome (Morales and Margolis 2017).

The substantia nigra is divided into the pars compacta containing dopaminergic neurons and the pars reticulata containing GABAergic neurons. The pars compacta appear dark in color because of high neuromelanin content, which is formed from a precursor molecule in the dopamine synthesis pathway. The pars compacta send dopaminergic projections to the striatum, within which the dopaminergic axons form synapses on two separate populations of neurons. Neurons expressing the D1-type dopamine receptor in the striatum send outputs to the internal segment of the globus pallidus forming the direct pathway of the basal ganglia. On the other hand, neurons expressing the D2-type dopamine receptor send outputs to the external segment of the globus pallidus, which inhibits excitatory neurons in the stimulatory subthalamic nucleus that projects to the internal segment of the globus pallidus, forming the indirect pathway (Figure 3). The neurons in the direct pathway of the striatum inhibit neurons in the internal segment of globus pallidus, whereas striatal neurons in the indirect pathway work to stimulate them. The internal segment of the globus pallidus sends inhibitory outputs to the thalamus and controls motor actions, which can be inhibition of unwanted motor output or release of motor action. The substantia nigra pars compacta have inputs from the pons, which receives inputs from afferent sensory neurons, the amygdala, the hypothalamus, etc. (Sonne et al. 2024). The neuronal activities of the substantia nigra pars compacta seem to be influenced by emotions.

Hippocampus

The hippocampus (Figure 2B of Chapter 4) plays a critical role in the formation of declarative memory as explained in chapter six. In addition, its involvement in processing emotions such as fear, anxiety and pleasure has been proposed (Figure 2) (Malezieux et al. 2023). In particular, the importance of the ventral or anterior (in primates) part of the hippocampus in processing emotions has been reported (Fanselow and Dong 2010). The hippocampus has bidirectional neuronal connection with the adjacent entorhinal cortex, the amygdala and the thalamus, and has a direct output to the medial prefrontal cortex (Jin and Maren 2015). The ventral and dorsal hippocampus show different neuronal connections, and bidirectional neuronal connections between the ventral hippocampus and the amygdala are noteworthy (Fanselow and Dong 2010).

The entorhinal cortex has bidirectional connections with diverse cerebral cortical areas, and send glutamatergic outputs to the dentate gyrus of the hippocampus. Granule neurons in the dentate gyrus send outputs to pyramidal neurons in the CA3 region, which in turn send outputs to pyramidal neurons in CA1 region of the hippocampus. All of these neurons are glutamatergic and form a major three neuron pathway in the hippocampus. CA1 pyramidal neurons send outputs to the entorhinal cortex forming a loop structure, although there are divergence and convergence of neuronal connections in the pathway, and there are also other neuronal pathways (Purves et al. 2019, Luo 2021).

Summary and Conclusion

Emotions and feelings are subjective conscious experiences, but they are also accompanied with some objective changes of body states such as the heart rate, blood pressure, facial expression, etc. An idea has been proposed that not only are emotions accompanied with body state changes but the latter also influence subjective emotions. Most emotions can be classified to positive ones such as pleasure, love and admiration, etc., and negative ones such as fear, anger, sadness, etc. In any case emotions are self-centered experiences.

Both fear and anxiety evoke defensive behavioral responses that work to avoid or reduce harm and seem to support the survival of an animal. Fear is caused by a factual and acute threat, whereas anxiety is evoked by potential and/or anticipated threat. The amygdala has attracted much attention as the center in the brain responsible for fear. However, following studies showed that the ventral hippocampus, the medial prefrontal cortex and the bed nucleus of the stria terminalis (BNST) play roles in the evaluation of threat, and that the central amygdala, the hypothalamus, the lateral septum, etc. are involved in the expression of fear responses.

Anger takes place when a goal of an individual is blocked, such as by being interfered by other individuals. Anger is not an avoidance-related but rather approach-related emotion, and anger sometimes elicits aggressive behaviors. Aggression seems to serve actions like securing resources, such as food, territory and/or a mating partner, and defending oneself and the family. Aggression costs a lot metabolically, and is potentially risky. The core neuronal regions processing aggression are the medial amygdala, BNST, the ventromedial hypothalamus, etc. The medial prefrontal cortex tonically suppresses the hypothalamus to block aggression, although the subpopulation of neurons there promote aggressive behaviors.

The original role of disgust has been considered to defend the animal's body from toxins and diseases, and the roles have been extended to defenses from various potential threats. Most humans experience disgust against people violating social norms and/or moral values. You may think that the above humans' behaviors are likely to cause anger rather than disgust. However, aggressive behaviors cost a lot, whereas avoidance reaction is cost saving. The involvement of the anterior insular cortex and the amygdala in toxicity- or disease-related disgust has been reported.

Pleasure is an internal reward for an animal. It occurs when an animal gets or experiences something good, which contributes to its survival or the production of its offsprings, and promotes behaviors which are likely to cause such outcomes. Previously, dopamine was considered to be crucial in experiencing pleasure, which has been questioned by recent studies. It has been proposed that there are separable processes regarding internal rewards. One is the motivation process for reward called wanting, and another is the core reaction to hedonic stimulation or pleasure itself called liking, and it has been considered that dopamine plays a role in wanting but not in liking. Thus, dopamine is suggested to strongly evoke a certain behavior towards getting rewards without experiencing pleasure. Dopamine has been also shown to be involved in reinforcement learning. Pleasure is often accompanied with specific facial expressions. Hot spots inducing facial hedonic responses by opioid application, have been found in the nucleus accumbens, the ventral pallidum, the orbitofrontal cortex and the insular cortex, which are functionally connected.

Subcortical regions such as the hypothalamus, the amygdala, and the nucleus accumbens, the cortical regions such as the insular, the anterior cingulate, the medial prefrontal and the orbitofrontal cortices, and the brain stem region PAG are involved in multiple emotions. There are strong interactions between emotions and interoceptions. Interoceptive information such as about the heart rate and breathing is transmitted to the insular cortex, and influences emotions such as fear and anxiety. On the other hand, emotions affect functions of various internal organs such as the gastrointestinal tract, the heart, the lung, the immune system, etc. Neurons working toward opposite directions are found in neighboring regions in the neuronal structures involved in emotions such as BNST processing anxiety, the medial frontal cortex controlling aggressive behavior and the nucleus accumbens implicated in hedonic responses. These neuronal organization together with the overlap of neuronal areas involved in processing multiple emotions, might cause unexpected interactions among emotions. The involvement of the medial prefrontal cortex in emotions imply strong influences of emotions to attention and decision making. It seems that emotions are processed and controlled by large-scale neuronal networks, including various loops and mutual interactions in broad brain regions.

References

Afifi, A.K. 2003. The basal ganglia: a neural network with more than motor function. Seminars Pediat. Neurol. 10: 3–10.

Anderson, D.J. 2016. Circuit modules linking internal states and social behaviour in flies and mice. Nat. Rev. Neurosci. 17: 692–704.

Arias, J.A., C. Williamsa, R. Raghvania, M. Aghajanic, S. Baezd, C. Belzunge et al. 2020. The neuroscience of sadness: A multidisciplinary synthesis and collaborative review. Neurosci. Biobehav. Rev. 111: 199–228.

Baliki, M.N. and A.V. Apkarian. 2015. Nociception, pain, negative moods, and behavior selection. Neuron 87: 474–491.

Benarroch, E.E. 2012. Periaqueductal gray: an interface for behavioral control. Neurol. 78: 210–217.

Battle, L. and D.L. Diab. 2022. Is envy always bad? An examination of benign and malicious envy in the workplace. Psychol. Rep. 332941221138476.

Berridge, K.C. and M.L. Kringelbach. 2015. Pleasure systems in the brain. Neuron 86: 646–664.

Bliss-Moreau, E., J.E. Toscano, M.D. Bauman, W.A. Mason and D.G. Amaral. 2011. Neonatal amygdala lesions alter responsiveness to objects in juvenile macaques. Neurosci. 178: 123–132.

Burkart, J.M., O. Allon, F. Amici, C. Fichtel, C. Finkenwirth, A. Heschl et al. 2014. The evolutionary origin of human hyper-cooperation. Nat. Commun. 5: 4747.

Cambridge Learner's Dictionary. 4th Ed. 2012 Cambridge Univ. Press, Cambridge, U.K.

Carlén, M. 2017. What constitutes the prefrontal cortex? Sci. 358: 478–482.

Caruana, F., A. Jezzini, B. Sbriscia-Fioretti, G. Rizzolatti and V. Gallese. 2010. Emotional and social behaviors elicited by electrical stimulation of the insula in the macaque monkey. Curr. Biol. 21: 195–199.

Carver, C.S. and E. Harmon-Jones. 2009. Anger is an approach related affect: evidence and implications. Psychol. Bull. 135: 183–204.

Catani, M., F. Dell'Acqua and M.T. de Schotten. 2013. A revised limbic system model for memory, emotion and behaviour. Neurosci. Biobehav. Rev. 37: 1724–1737.

Chapman, H.A. and A.K. Anderson. 2012. Understanding disgust. Ann. N. Y. Acad. Sci. 1251: 62–76.

Chapman, H.A., D.A. Kim, J.M. Susskind and A.K. Anderson. 2009. In bad taste: evidence for the oral origins of moral disgust. Sci. 323: 1222–1226.

Collins, K.L., H.G. Russell, P.J. Schumacher, K.E. Robinson-Freeman, E.C. O'Conor, K.D. Gibney et al. 2018. A review of current theories and treatments for phantom limb pain. J. Clin. Invest. 128: 2168–2176.

Connolly, T. and M. Zeelenberg. 2002. Regret in decision making. Curr. Direct. Psychol. Sci. 11: https://doi-org.kyoto-u.idm.oclc.org/10.1111/1467-8721.00203.

Critchley, H.D. and N.A. Harrison. 2013. Visceral influences on brain and behavior. Neuron 77: 624–638.

Damasio, A. and G.B. Carvalho. 2013. The nature of feelings: evolutionary and neurobiological origins. Nat. Rev. Neurosci. 14: 143–152.

Davis, M., D.M. Walker, D.L.L. Miles and C. Grillon. 2010. Phasic versus sustained fear in rats and humans: role of the extended amygdala in fear versus anxiety. Neuropsychopharmacol. 35: 105–135.

deWaal, F.B.M. and S.D. Preston. 2017. Mammalian empathy: behavioural manifestations and neural basis. Nat. Rev. Neurosci. 18: 498–509.

Dolensek, N., D.A. Gehrlach, A.S. Klein and N. Gogolla. 2020. Facial expressions of emotion states and their neuronal correlates in mice. Sci. 368: 89–94.

Dutton, D.G. and A.P. Aron. 1974. Some evidence for heightened sexual attraction under conditions of high anxiety. J. Personality Social Psychol. 30: 510–517.

Fanselow, M.S. and H. Dong. 2010. Are the dorsal and ventral hippocampus functionally distinct structures? Neuron 65: 7–19.

Feinstein, J.S., R. Adolphs, A. Damasio and D. Tranel. 2011. The human amygdala and the induction and experience of fear. Curr. Biol. 21: 34–38.

Feinstein, J.S., C. Buzza, R. Hurlemann, R.L. Follmer, N.S. Dahdaleh, W.H. Coryell et al. 2013. Fear and panic in humans with bilateral amygdala damage. Nat. Neurosci. 16: 270–272

Feldman, R. 2017. The neurobiology of human attachments. Trends Cogn. Sci. 21: 80–99.

Fong, H., J. Zheng and D. Kurrasch. 2023. The structural and functional complexity of the integrative hypothalamus. Sci. 382: 388–394.

Floresco, S.B. 2015. The nucleus accumbens: an interface between cognition, emotion, and action. Annu. Rev. Psychol. 66: 25–52.

Gehrlach, D.A., N. Dolensek, A.S. Klein, R.R. Chowdhury, A. Matthys, M. Junghänel et al. 2019. Aversive state processing in the posterior insular cortex. Nat. Neurosci. 22: 1424–1437.

Georgescu, I.A., D. Popa and L. Zagrean. 2020. The anatomical and functional heterogeneity of the mediodorsal thalamus. Brain Sci. 10: 624.

Gilam, G., J.J. Gross, T.D. Wager, F.J. Keefe and S.C. Mackey. 2020. What is the relationship between pain and motion? Bridging constructs and communities. Neuron 107: 17–21.

Han, Q., S. Quadflieg and C.J.H. Ludwig. 2023. Decision avoidance and post-decision regret: a systematic review and meta-analysis. PLoS One 18: e0292857.

Harris, C.R. and C. Prouvost. 2014. Jealousy in dogs. PLoS One 9: e94597.

Hernandez, G., H. Rajabi, J. Stewart, A. Arvanitogiannis and P. Shizgal. 2008. Dopamine tone increases similarly during predictable and unpredictable administration of rewarding brain stimulation at short inter-train intervals. Behav. Brain Res. 188: 227–232.

Hernandez-Lallement, J., A.T. Attah, E. Soyman, C.M. Pinhal, V. Gazzola and C. Keysers. 2020. Harm to others acts as a negative reinforcer in rats. Curr. Biol. 30: 949–961.

Hess, W.R. and K. Akert. 1955. Experimental data on role of hypothalamus in mechanism of emotional behavior. AMA Arch. Neurol. Psychiatry 73: 127–129.

Hu, H. 2016. Reward and aversion. Annu. Rev. Neurosci. 39: 297–324.

Janak, P.H. and K.M. Tye. 2015. From circuits to behaviour in the amygdala. Nat. 517: 284–292.

Jin, J. and S. Maren. 2015. Prefrontal-hippocampal interactions in memory and emotion. Front. Syst. Neurosci. 9: 170.

Khouri, L. and I. Nelken. 2015. Detecting the unexpected. Curr. Opin. Neurobiol. 35: 142–147

Koch, M. 1999. Neurobiology of startle. Prog. Neurobiol. 59: 107–128.

Lamm, C., M. Rütgen and I.C. Wagner. 2019. Imaging empathy and prosocial emotions. Neurosc. Lett. 693: 49–53.

Lebow, M.A. and A. Chen. 2016. Overshadowed by the amygdala: the bed nucleus of the stria terminalis emerges as key to psychiatric disorders. Mol. Psychiat. 21: 450–463.

LeDoux, J. 2007. The amygdala. Cur. Biol. 17: R868–R874.

Lischinsky, J.E. and D. Lin. 2020. Neural mechanisms of aggression across species. Nat. Neurosci. 23: 1317–1328.
Luo, L. 2021. Principles of Neurobiology. 2nd Ed. CRC Press, Boca Raton, FL.
Malezieux, M., A.S. Klein and N. Gogolla. 2023. Neural circuits for emotion. Annu. Rev. Neurosci. 46: 211–31.
Mellers, B., K. Fincher, C. Drummond and M. Bigony. 2013. Surprise: a belief or an emotion? Prog. Brain Res. 202: 3–19.
Mendez, M.F. 2023. A brain mechanism for hate. J. Neuropsychiatry. Clin. Neurosci. 35: 262–263.
Morales, M. and E.B. Margolis. 2017. Ventral tegmental area: cellular heterogeneity, connectivity and behaviour. Nat. Rev. Neurosci. 18: 73–85.
Olds, J. 1956. Pleasure centers in the brain. Sci. Am. 195: 105–116.
Olds, J. and P. Milner. 1954. Positive reinforcement produced by electrical stimulation of septal area and other regions of rat brain. J. Comp. Physiol. Psychol. 47: 419–427.
Panksepp, J. 2011. Cross-species affective neuroscience decoding of the primal affective experiences of humans and related animals. PLoS One 6: e21236.
Paradiso, E., V. Gazzola and C. Keysers. 2021. Neural mechanisms necessary for empathy-related phenomena across species. Curr. Opin. Neurobiol. 68: 107–115.
Pecina, S. and K.C. Berridge. 2005. Hedonic hot spot in nucleus accumbens shell: where do mu-opioids cause increased hedonic impact of sweetness? J. Neurosci. 25: 11777–11786.
Porreca, F. and E. Navratilova. 2017. Reward, motivation and emotion of pain and its relief. Pain 158(Suppl 1): S43–S49.
Pessoa, L. 2017. A network model of the emotional brain. Trends Cogn. Sci. 21: 357–371.
Purves, D., G.J. Augustine, D. Fitzpatrick, W.C. Hall, A.S. LaMantia, R.D. Mooney et al. 2019. Neuroscience. 6th Ed. Sinauer, Sunderland, MA.
Richard, R., N. Tazi, D. Frydecka, M.S. Hamid and A.A. Moustafa. 2022. A systematic review of neural, cognitive, and clinical studies of anger and aggression. Curr. Psychol. doi.org/10.1007/s12144-022-03143-6.
Rolls, E.T. 2015. Limbic systems for emotion and for memory, but no single limbic system. Cortex 62: 119–157.
Rolls, E.T. 2019. The cingulate cortex and limbic systems for emotion, action, and memory. Brain Struct. Funct. 224: 3001–3018.
Ross, H.E., S.M. Freeman, L.L. Spiegel, X. Ren, E.F. Terwilliger and L.J. Young. 2009. Variation in oxytocin receptor density in the nucleus accumbens has differential effects on affiliative behaviors in monogamous and polygamous voles. J. Neurosci. 29: 1312–1318.
Rudebeck, P.E. and E.L. Rich. 2018. Orbitofrontal cortex. Curr Biol. 28: 1075–R1095.
Schultz, W. 1998. Predictive reward signal of dopamine neurons. J. Neurophysiol. 80: 1–27.
Sherman, S.M. 2001. Tonic and burst firing: dual modes of thalamocortical relay. Trends Neurosci. 24: 122–126.
Sherman, E.R., J.J. Thomas and E.N. Cahill. 2022. The bed nucleus of the stria terminalis in threat detection: task choice and rodent experience. Emerg. Top. Life Sci. 6: 457–466.
Sienkiewicz-Jarosz, H., A. Scinska, W. Kuran, D. Ryglewicz, A. Rogowski, E. Wrobel et al. 2005. Taste responses in patients with Parkinson's disease. J. Neurol. Neurosurg. Psychiatry 76: 40–46.
Sonne, J., V. Reddy and M.R. Beato. 2024. Neuroanatomy, substantia nigra. StatPearls [Internet] PMID: 30725680.
Susskind, J.M., D.H. Lee, A. Cusi, R. Feiman, W. Grabski and A.K. Anderson. 2008. Expressing fear enhances sensory acquisition. Nat. Neurosci. 11: 843–850.
Takahashi, H., M. Kato, M. Matsuura, D. Mobbs, T. Suhara and Y. Okubo. 2009. When your gain is my pain and your pain is my gain: neural correlates of envy and schadenfreude. Sci. 323: 937–939.
Tovote. P., J.P. Fadok and A. Lüthi. 2015. Neuronal circuits for fear and anxiety. Nat. Rev. Neurosci. 16: 317–331.
Uddin, L.Q., J.S. Nomi, B. Hebert-Seropian, J. Ghaziri and O. Boucher. 2017. Structure and function of the human insula. J. Clin. Neurophysiol. 34: 300–306.
Venkatraman, A., B.L. Edlow and M.H. Immordino-Yang. 2017. The brainstem in emotion: a review. Front. Neuroanat. 11: 15.

Vogt, B.A. 2005. Pain and emotion interactions in subregions of the cingulate gyrus. Nat. Rev. Neurosci. 6: 533–544.

Wallis, J.D. 2007. Orbitofrontal cortex and its contribution to decision-making. Annu. Rev. Neurosci. 30: 31–56.

Yan, H., N.A. Shlobin, Y. Jung, K.K. Zhang, N. Warsi, A.V. Kulkarni et al. 2023. Nucleus accumbens: a systematic review of neural circuitry and clinical studies in healthy and pathological states. J. Neurosurg. 138: 337–346.

Zeki, S. 2007. The neurobiology of love. FEBS Lett. 581: 2575–2579.

Zeki, S. and J.P. Romaya. 2008. Neural correlates of hate. PLoS ONE 3: e3556.

Zhang, R., H. Deng and X. Xiao. 2024. The insular cortex: an interface between sensation, emotion and cognition. Neurosci. Bull. https://doi.org/10.1007/s12264-024-01211-4.

9

Neuromodulators

Neuromodulator Molecules

Neuromodulation can be defined as any communications between neurons, or neurons and non-neuronal cells such as glial cells, which are caused by a released chemical that acts in a way not fast, not point-to-point, and/or not simply by excitation or inhibition. Neuromodulator molecules often influence global behavioral states such as the effect of serotonin on mood and the effect of norepinephrine on sleep and wakefulness, etc. (Lee and Dan 2012). Neuromodulators can alter the working states of hard-wired neuronal circuits so that they adapt to the changing status of an animal in the environment (Bucher and Marder 2013), and play crucial roles in emotions. They are often released diffusely in a paracrine fashion, which is called volume transmission. The effects of a neuromodulator on neurons are not confined to being either excitatory or inhibitory; they are diverse depending on the receptors and are often relatively long-lasting, because receptors are GPCRs in most cases. There are various kinds of neuromodulators, and most neurons respond to multiple neuromodulators. Amines such as dopamine, norepinephrine, serotonin, histamine, etc., peptides such as enkephalins, oxytocin, vasopressin, etc., lipid-like molecules such as endocannabinoids and steroids, and also gasses such as nitric oxide are regarded neuromodulators. In the following sections, I explain the roles and functions of the respective neuromodulator molecules.

Dopamine

Dopamine (Figure 4 of Chapter 3) is mainly synthesized from an amino-acid tyrosine in the ventral tegmental area, the substantia nigra pars compacta in the midbrain, etc. (Figure 1 of Chapter 8) (Klein et al. 2019, Speranza et al. 2021). There are two types of dopamine releases; they are phasic and tonic releases (Floresco et al. 2003). Burst firings of neurons in the ventral tegmental area induce massive and phasic dopamine release to synaptic clefts in the nucleus accumbens, and the released dopamine are rapidly removed by reuptake. On the other hand, increased population activities of neurons without bursting in the ventral tegmental area, cause tonic increase in dopamine concentration in the extra-synaptic space, which are less influenced by

DOI: 10.1201/9781003518822-9

reuptake. There are five types of dopamine receptors, which are named D_1–D_5. They are classified into D_1-type consisting of D_1 and D_5 receptors, and D_2-type consisting of D_2, D_3 and D_4 receptors. D_2-type receptors have 10–100 times higher affinities than D_1-type receptors, and are activated by a tonic level of dopamine. D_1-type receptors are mainly coupled to Gs protein and activate adenylyl cyclase producing cAMP, whereas D_2-type receptors are coupled to Gi protein inhibiting adenylyl cyclase (Klein et al. 2019). It should be noted that some dopaminergic neurons co-release other neurotransmitters such as glutamate or GABA as explained in chapters three and eight, implying that they have multiple functional roles in information transmission (Lerner et al. 2021).

Dopaminergic neurons in the substantia nigra pars compacta send outputs to the dorsal striatum, and play roles in motor control, etc. Dopaminergic control of body movements is performed through two pathways, direct and indirect pathways as explained in chapter eight (Figure 3 of Chapter 8) (Purves et al. 2019, Luo 2021). There are two types of intermingled GABAergic inhibitory projection neurons in the dorsal striatum, which are mostly silent at rest and receive glutamatergic inputs from the cerebral cortices and the thalamus. One type are neurons expressing D_1 receptor constituting the direct pathway, which send inhibitory outputs to neurons in the globus pallidus internal segment in the basal ganglia. The latter neurons are intrinsically active and send tonic inhibitory outputs to the thalamus and the brainstem. The above route forms disinhibitory pathways consisting of intrinsically silent inhibitory neurons followed by intrinsically active inhibitory neurons, and the net outputs are excitatory and work to initiate motion. Another type of GABAergic neurons in the dorsal striatum are neurons expressing D_2 receptor constituting indirect pathways, which send inhibitory outputs to GABAergic neurons in the globus pallidus external segment, which in turn send inhibitory outputs to neurons in the globus pallidus internal segment directly or indirectly through the subthalamic nucleus. In this route there are three sequentially connected inhibitory neurons, and the net outputs to the thalamus and the brainstem are inhibitory. The activation of the D_1 receptor results in depolarization of neurons, enhancing the information flow in the direct pathway; whereas the activation of the D_2 receptor results in hyperpolarization of neurons suppressing the indirect pathway.

Dopaminergic neurons in the ventral tegmental area send axons to the prefrontal cortex, the cingulate cortex, the nucleus accumbens, the amygdala, the hippocampus, etc. (Figures 1 and 2 of Chapter 8), and play roles in the control of motivation, reward seeking, attention, reinforcement learning, etc. (Lerner et al. 2021, Thiele and Bellgrove 2018). The ventral tegmental area is implicated in positive rewards and appetite-motivated behaviors. It plays a crucial role in wanting of pleasure. However, it is to be noted that aversive stimuli and stress also activate dopaminergic neurons in the ventral tegmental area as explained in chapter eight (Bromberg-Martin et al. 2010, Klein et al. 2019). In reinforcement learning dopamine neurons in the ventral tegmental area have been considered to code a reward prediction error and to enhance the occurrence of synaptic plasticity, which promotes reinforcement learning as explained in chapter six (Schultz et al. 1997, Lerner et al. 2021, Luo 2021, Speranza et al. 2021).

Malfunctions of dopaminergic systems have been related to various neurological and neuropsychiatric diseases (Lerner et al. 2021). Parkinson's disease is caused by the loss of dopaminergic neurons in the substantia nigra pars compacta. The patients show motor symptoms such as difficulty to start motion, slow movement, tremor, rigidity, etc., and also cognitive impairment and depression. The implication of abnormalities of the dopaminergic system in schizophrenia and attention deficit hyperactivity disorders has also been suggested. Various drugs targeting dopamine receptors, the D_2 receptor in particular, have been used for therapy in Parkinson's disease, schizophrenia, etc. In addition, the potential use of stem cell-derived dopamine neurons for the therapy of Parkinson's disease has been studied. Such brain disorders are explained further in chapter eleven.

Norepinephrine

Norepinephrine (Figure 4 of Chapter 3) is converted from dopamine by an enzyme named dopamine-β-hydroxylase in neurons of the locus coeruleus (Figure 1 of Chapter 8), a small nucleus located in the brainstem. About 80% of neurons in the locus coeruleus also express neuropeptide galanin and a smaller population of neurons express neuropeptide Y. Outputs from the locus coeruleus extend to nearly all regions of the brain except the striatum, the nucleus accumbens, the globus pallidus and the substantia nigra, all of which receive abundant dopaminergic projections. Thus, there is some division of functional roles between norepinephrine and dopamine (Schwarz and Luo 2015, Luo 2021). Norepinephrine is also called noradrenaline, and its receptor is usually called adrenergic receptors, which can be activated not only by norepinephrine but also by epinephrine. Adrenergic receptors are classified into 3 subtypes, α_1-, α_2- and β-adrenergic receptors. All of them are coupled to heterotrimeric G proteins. α_1-AR is mainly coupled to Gq protein, which activates phospholipase C producing IP3 and diacylglycerol. β-AR is primarily coupled to Gs protein, which activates adenylyl cyclase producing cAMP, whereas α_2-AR is coupled to Gi protein, which suppresses adenylyl cyclase. However, other intracellular signaling pathways are also likely to be activated by each subtype (Hirano and Inoshita 2021).

Norepinephrine is implicated in the control of arousal, response to stress, attention, learning, etc. (Schwarz and Luo 2015). The action potential firing rates of the locus coeruleus neurons are high in arousal states, and their optogenetic activation promotes sleep to wake transition (Carter et al. 2010). It was also reported that their high-frequency tonic activation during stress-induced behavior could cause anxiety-like and aversive behaviors, and that their inhibition during stressful stimuli suppressed anxiety-like behaviors (McCall et al. 2015). In the sensory cortex there are α_1, α_2 and β adrenergic receptors. The activation of α_1 or β receptor causes excitation, and that of α_2 receptor causes inhibition in general, and the overall effects of norepinephrine are the decrease in spontaneous activities and the increase in activities driven by a salient input (Thiele and Bellgrove 2018). Such effects of norepinephrine seem to improve the signal to noise ratio of input signals and strengthen the activities related to salient or attended objects. Norepinephrine in the prefrontal cortex has been considered to play important roles in top-down attention (Thiele and Bellgrove 2018).

Norepinephrine also plays a role in the regulation of synaptic plasticity and learning. Norepinephrine facilitates memory formation, consolidation, maintenance and retrieval in general (Sara 2009, Tully and Bolshakov 2010). Previous studies suggested that when an animal encounters a novel object, arousal occurs and neurons in the locus coeruleus are activated. Norepinephrine released in the hippocampus and the amygdala facilitates LTP and memory formation through β-adrenergic receptors. Norepinephrine also facilitates induction of cerebellar LTD and contributes to the cerebellum-dependent motor learning paradigms (Hirano and Inoshita 2021). These effects might underly enhancement of learning by emotions.

Epinephrine or adrenaline is converted from norepinephrine and binds to adrenergic receptors. Dopamine, norepinephrine and epinephrine are structurally related and classified to catechol amines. Epinephrine is well known as a hormone secreted from the adrenal gland, and plays crucial roles in stress responses, in emergency or in the excited state as explained in chapter four. It increases the heart rate, the blood pressure, the blood glucose level, and the opening of the pupils, etc. Epinephrine is also produced in small brain regions in the medulla oblongata, and the neurons there send axons to the hypothalamus, the spinal cord, etc.

The malfunction of the norepinephrine system has been considered to be involved in various psychiatric and neurological disorders such as anxiety disorders, depression, Parkinson's disease, etc. Reuptake blockers of norepinephrine are sometimes prescribed to patients claiming anxiety and/or depression, although they show some side effects (Schwarz and Luo 2015).

Serotonin

Serotonin or 5-hydroxytryptamine (5-HT) is a type of indole amines and structurally distinct from catechol amines such as dopamine, norepinephrine and epinephrine (Figure 4 of Chapter 3). It is converted from an amino acid tryptophane. Serotonin is found in various organs in the body such as the brain, the lung, the kidney, the blood, the gastrointestinal tract, etc. Actually about 95% of serotonin is found outside of the brain, in the enterochromaffin cells in the gastrointestinal tract and platelets in the blood in particular. In the brain, serotonin is mainly synthesized in nine groups of neurons, called the raphe nuclei in the midbrain and in the pons (Figure 1 of Chapter 8). They send outputs to various regions of the brain and also to the spinal cord (Mohammad-Zadeh et al. 2008, Berger et al. 2009). Serotonin mainly works in volume transmission (Hensler 2006). There are 7 types of serotonin receptors named 5-HT_{1-7}. Among them 5-HT_3 is a chemically-gated ion channel, and the others are GPCRs. 5-HT_1 and 5-HT_5 are coupled to Gi protein, 5-HT_2 is coupled to Gq protein, and 5-$HT_{4, 6 \text{ and } 7}$ are coupled to Gs protein (Nichols and Nichols 2008).

Serotonin has been considered to be involved in various brain functions such as control of sleep cycles, anxiety, nociception, appetite, etc. (Mohammad-Zadeh et al. 2008). The malfunction of serotonin system has been implicated in depression, anxiety disorders, schizophrenia, etc. Selective serotonin reuptake inhibitors have been widely used as effective antidepressants with limited side effects. Selective serotonin reuptake inhibitors are also prescribed to patients suffering from anxiety disorders. Other drugs targeting serotonin system such as 5-HT_{1A} antagonists have also been used sometimes (Hensler 2006, Cowan 2008, Mohammad-Zadeh et al. 2008).

Histamine

Histamine is a type of imidazole amines and is synthesized from an amino acid histidine through oxidative decarboxylation (Figure 3 of Chapter 4). Histamine is found not only in the brain but also in the gut, the skin and the immune system. Histamine plays critical roles in allergy, inflammation, etc. in peripheral regions of the body. In the brain, histaminergic neurons are located in the tuberomamillary nucleus in the posterior hypothalamus. They project to various areas of the brain and the spinal cord. Some histaminergic neurons express peptide neuromodulators, and GABA is also a co-transmitter of some histaminergic neurons. There are four types of histamine receptors named H_1–H_4 which are GPCRs. H_1 receptor is mainly coupled to Gq protein, and H_2 receptor to Gs protein. H_1 and H_2 receptors show mostly excitatory effects. H_3 receptor is mainly coupled to Gi protein and mediates inhibitory effects. The H_4 receptor is found mostly in peripheral tissues (Haas et al. 2008).

Histamine is implicated in sleep-wake regulation, circadian and feeding rhythms, thermoregulation, pain and itch, immunity, learning, etc. Drugs targeting histamine systems have been used for the treatment of allergy and pain including migraine. The side effects of these drugs are to make people sleepy. Histaminergic neurons are active only during waking hours. The activation of the H_1 receptor promotes wakefulness, and drugs that block H_1 receptor are the most commonly used medications for insomnia (Haas et al. 2008, Scammell et al. 2019).

Acetylcholine

Acetylcholine (Figure 4 of Chapter 3) is an important neurotransmitter working at neuromuscular synapses and in the parasympathetic systems. It also plays a role as a neuromodulator in the brain. Acetylcholine containing or cholinergic neurons are found in the tegmental nuclei, the basal forebrain including the medial septum, etc., and project into wide areas of the brain. There are both metabotropic and ionotropic acetylcholine receptors in the brain. The metabotropic receptors are coupled to either Gq proteins (M_1, M_3, and M_5 subtypes) or Gi/o proteins (M_2 and M_4 subtypes), located in both pre- and postsynaptic regions of neurons. Ionotropic acetylcholine receptors are also found not only on the postsynaptic membrane but also on other surface membranes of neurons including the presynaptic membrane, and seem to play modulatory roles on synaptic and/or neuronal activities together with metabotropic receptors. Acetylcholine seems to be involved in both synaptic and volume transmissions. Cholinergic systems have been implicated in arousal, attention, self-organization of cerebral cortical circuits, learning, control of autonomic functions, stress responses, etc. (Picciotto et al. 2012).

Enkephalin, Dynorphin and Endorphin

Opioids such as morphine have been used for more than thousands of years for pain treatment, feelings of euphoria, etc. Such a situation has made researchers think that there are some endogenous opioid-like substances, and attempts to find them have been carried out. In 1976, β-endorphin was identified as an endogenous substance possessing opioid functions, followed by the discovery of enkephalins and dynorphins. Enkephalins, dynorphins and endorphins are peptides and processed from proenkephalin, prodynorphin and proopiomelanocortin, respectively. Each of them has subtypes, and there are more than twenty opioid peptides. Enkephalins and dynorphins are widely distributed in the brain, the basal ganglia in particular. Some neurons in various brain areas including the hypothalamus, PAG, the hippocampus, etc. (Figure 1 of Chapter 8) seem to express these neuropeptides (van den Pol 2012, Benarroch 2012). The mRNAs encoding proenkephalin, prodynorphin and proopiomelanocortin are also highly expressed in some peripheral tissues such as the adrenal gland, the prostate, the testis, the ovary, the pituitary gland, the immune systems, etc. (Fricker et al. 2020, Che and Roth 2023).

Opioid receptors are classified as three types, μ, δ and κ opioid receptors. They are all GPCRs coupling to Gi/o proteins, and exert inhibitory effects on neurons (Terenius and Johansson 2010, Che and Roth 2023). Enkephalins are found in the presynaptic terminals of interneurons forming synapses on the neurons in the dorsal horn of the spinal cord which send pain signals to the brain (Terenius and Johansson 2010). As described in chapter eight, the microinjection of an opioid into restricted hot spots in the nucleus accumbens enhances hedonic reaction to sweet taste in rats (Pecina and Berridge 2005). These results together with other findings support the roles of opioids in pain suppression and liking of pleasure as explained in chapter eight (Fricker et al. 2020). One serious problem about opiates is that they cause addiction, which is explained in a later section of this chapter.

Oxytocin

Oxytocin is a peptide hormone synthesized in neurons of some nuclei in the hypothalamus, which send axons to the posterior pituitary gland and secrete oxytocin to blood vessels (Figure 2 of Chapter 4). The important roles of oxytocin in a female body are uterine contraction during parturition and milk secretion during lactation. In the brain, oxytocin neurons show both synaptic release and volume transmission, and co-release glutamate. The oxytocin receptor is the only one type GPCR coupled to Gi or Gq protein (Grinevich and Neumann 2021, Marsh et al. 2021).

In the brain oxytocin neurons project to reward-related and fear-related regions such as the nucleus accumbens, the ventral tegmental area, the central amygdala, BNST, etc. (Figure 1 of Chapter 8). In relation to these projection patterns, oxytocin shows anxiolytic and antistress effects, and contributes to increasing sociality including trust, empathy and altruism, although in some cases oxytocin may elicit defensive and aggressive responses (Marsh et al. 2021). It was reported that the optogenetic stimulation of some axons of oxytocin neurons in the central amygdala decreased freezing responses in fear-conditioned rats (Knobloch et al. 2012). The oxytocin neurons also extend axons to the prefrontal cortex, the cingulate cortex, the substantia nigra, the locus coeruleus, the dorsal raphe nucleus, etc. (Grinevich and Neumann 2021).

Cannabinoids

Cannabis, including marijuana, derived from the plant *Cannabis sativa,* has been used to obtain euphoric effects, therapeutic effects, etc. for about 1,000 years, although the use of cannabis is now illegal in many countries because of its addictive effects, etc. Δ^9-tetrahydrocannabinol (THC) is the major psychotropic component of cannabis, and its receptors in the brain are identified as CB1 and CB2 receptors. Both CB1 and CB2 receptors are GPCRs mainly coupled to Gi protein. The CB1 receptor is enriched in the brain, but it is also present in peripheral organs such as the liver, the adipose tissue and the skin. CB2 receptor is highly expressed in the immune system including the microglial cells in the brain. The search for endogenous substances binding to CB1 and CB2 receptors identified anandamide and 2-arachidonoylglycerol (2-AG), which are called endocannabinoids (Cristino et al. 2020, Lu and Mackie 2021).

Anandamide and 2-AG are synthesized from phospholipids on the cell membrane when neurons are activated. The intracellular increase in calcium ions caused by depolarization and/or GPCR activation trigger synthesis. In neurons, CB1 receptor is particularly enriched on presynaptic terminals, and works to suppress synaptic transmission. Endocannabinoid, produced in the activated postsynaptic neuron, travels retrogradely to a presynaptic terminal and binds to CB1 receptor. The binding results in the suppression of release of excitatory or inhibitory neurotransmitter, which are called depolarization-stimulated suppression of excitation or depolarization-stimulated suppression of inhibition, respectively. Endocannabinoids also show other effects on neuronal activities (Lu and Mackie 2021). It is to be noted that there are endocannabinoids other than anandamide and 2-AG, and that there are also cannabinoid receptors other than CB1 and CB2 receptors.

Steroids

Steroids are a type of lipid synthesized from cholesterol, and several of them including cortisol, aldosterone, progesterone, estrogens, androgens are produced in the endocrine or reproductive organs and work as hormones. Such steroid hormones permeate through the cellular membrane in general, bind to their receptors inside a cell, go into a nucleus and exert effects through the alteration of the gene transcription (Alberts et al. 2023). The above processes usually take hours.

The synthesis of some steroid molecules also takes place in the brain, and they show faster modulatory effects on neuronal activities (Remage-Healey 2014, Rudolph et al. 2016, Joëls 2018). Fast neuro-modulatory actions of steroids are mediated through GPCRs such as a metabotropic glutamate receptor (Remage-Healey 2014, Rudolph et al. 2016), or through the action on ionotropic GABA receptors (Belelli 2021). Endogenous metabolites of progesterone and deoxycorticosterone which are produced in the brain, are positive allosteric modulators of ionotropic GABA receptors and enhance their functions. Thus, some steroids exhibit anxiolytic, anticonvulsant, analgesic and sedative effects.

Nitric Oxide

Nitric oxide (NO) is a gaseous molecule diffusing freely through the cell membrane of the tissue. NO is mainly synthesized by an enzyme named NO synthases (NOS), which oxidizes L-arginine (L-Arg) to citrulline and NO. There are four types of NO synthases, endothelial, neuronal, inducible and mitochondrial NO synthases. In the periphery, the important function of NO is the relaxation of vascular smooth muscle cells. Thus, an NO producing chemical, such as nitroglycerin, is used for the treatment of angina pectoris. NO binds to the guanylyl cyclase in a cell and activates it, resulting in the alteration of the metabolic state of the cell (Picón-Pagès et al. 2018, Alberts et al. 2023).

In the brain, NO is mostly produced by neuronal NO synthase in postsynaptic neurons in a NMDA-type glutamate receptor-dependent manner, and works as a retrograde messenger. NO has been implicated in the regulation of synaptic plasticity in both pre- and postsynaptic neurons (Picón-Pagès et al. 2018).

Drugs of Abuse and Addiction

Opioids, cannabinoids and their related substances are subject to abuse and may cause addiction (Anghel et al. 2023). Opioids, such as heroin, etc., which is produced from morphine, induces euphoria and sedation, that is a drowsy state with reduced arousal. They also cause respiratory depression, etc. in the body. Thus, they are classified as depressants. A prolonged use of opioids disrupts the endogenous opioid system or the reward system, and causes anxiety, depression, dysphoria, etc. Opioids seem to induce euphoria through their actions in the nucleus accumbens as explained in a previous section and in chapter eight. In addition, opioids increase the extracellular dopamine level through δ- and μ-opioid receptors in the nucleus accumbens, which might cause wanting of pleasure and addiction (Saigusa et al. 2017).

Cocaine, an alkaloid derived from the leaves of the plants *Erythroxylum coca* et al., amphetamine, methamphetamine, methylenedioxy-methamphetamine (MDMA), etc. induces euphoria, increase alertness and reduces fatigue. Thus, they are classified as stimulants. Cocaine, methamphetamine and MDMA increase the concentration of catecholamines including dopamine and norepinephrine in the extracellular space, and seem to cause wanting and also alerting through dopamine and norepinephrine systems (Anghel et al. 2023).

Cannabis such as marijuana affects the endocannabinoid system through interaction with the presynaptic CB1 receptor on inhibitory GABAergic interneurons in the reward pathway, and seems to cause euphoria, relaxation, etc. The interaction reduces GABA release, causing the disinhibition of dopaminergic neurons, and increases the dopamine level (Anghel et al. 2023). Thus, dopamine signaling seems to be critically involved in addiction.

Alcohol is a depressant and can also cause addiction through enhancement of the GABAergic system, etc. Alcohol induces dose-dependent sedation, although there is a stimulating phase with relaxation, loss of inhibition, emotional arousal and pleasure in moderate consumption. At a higher dose, depression, sadness, anger, aggression, etc. may occur (Anghel et al. 2023).

Physical Neuromodulation

Neuronal activities in certain brain areas can be altered not only by neuromodulator molecules but also by various physical and/or chemical stimulation methods. The definition of neuromodulation by the International Neuromodulation Society is "the alteration of nerve activity through the targeted delivery of a stimulus, such as an electrical stimulation or chemical agents, to specific neurological sites in the body" (Knotkova et al. 2021). Optogenetic stimulations, electrical stimulations, magnetic stimulations and ultrasound stimulations of the brain regions have been used for research and for therapeutic purposes of neuromodulation.

Optogenetic stimulations have been widely used in animal experiments, because the methods allow activation or inactivation of specific neurons at desired timings. In optogenetics, the optically-activatable ion channel is expressed in a certain type of neurons using a cell-specific promoter, and the channel is activated by applying light with a specific wave length. Thus, the target specification of the stimulation is not confined only to a region, but also to a type of neurons such as those expressing a particular type of receptor, etc. However, the method is invasive in two respects. One is artificial overexpression of an exogenous protein using an exogenous vector, and the other is setting up of the light application equipment in or on the brain. Thus, optogenetic modulation has been confined to animal experimentations, although attempts or research to apply this technique to human therapies have been on going (Henderson et al. 2009).

Electrical stimulations such as peripheral nerve stimulation, transcutaneous electrical nerve stimulation, transcranial direct current stimulation and deep brain stimulation, also transcranial magnetic stimulation and transcranial ultrasound stimulation, have been used for treatments of Parkinson's disease, essential tremor, dystonia, chronic pain, etc. in humans (Knotkova et al. 2021, Krauss et al. 2021, Darmani et al. 2022). Deep brain electrical stimulation using implanted electrodes is a standard care in Parkinson's disease, essential tremor and dystonia (Darmani et al. 2022). The electrodes are positioned to target various nodes along the basal ganglia-thalamus neuronal circuits. The stimulation of the thalamus is most efficient for tremors; that of the pallidum in the basal ganglia is efficient for rigidity and dyskinesias; whereas that of the subthalamic nucleus can treat tremors, akinesia, rigidity and dyskinesia (Hariz and Blomstedt 2022).

There are non-invasive methods for neuromodulation in humans. Transcranial direct current stimulation uses a device that provides low intensity electrical current to the surface of the head, which results in subthreshold modulation of neuronal excitability. In transcranial magnetic stimulation. a strong magnetic field is produced by the electromagnetic coils. The magnetic field gives rise to an electrical current in the brain, which modulates neuronal excitability. Transcranial ultrasound stimulation is a non-invasive method that can focus on both cortical and deep brain targets with a high spatial resolution of a few cubic millimeters. It has been considered that the ultrasound waves influence the neuronal cell membrane and/or affect mechanosensitive ion channels, and change excitabilities of neurons. The strength of the stimulation by ultrasound seems to be relatively weak as compared with transcranial magnetic stimulation (Darmani et al. 2022).

All of the above physical neuromodulation methods applied to humans may cause side effects, in some cases, partly due to the spread of stimuli outside of the target neurons or region. An alternative therapy method to use induced pluripotent stem cells (iPSCs) for Parkinson's disease has been studied. Parkinson's disease is a neurodegenerative disease caused by the progressive loss of dopaminergic neurons in the substantia nigra pars compacta. iPSCs are capable of differentiating into various cells in the body including dopaminergic neurons. Thus, attempt to implant iPSCs derived dopaminergic neurons into the brain has been considered as a potential therapy of Parkinson's disease (Shastry et al. 2023). Neurodegenerative diseases including Parkinson's disease are explained further in chapter eleven.

Summary and Conclusion

Neuromodulation can be defined as communication mainly between neurons caused by a released chemical. Neuromodulator molecules can alter the working states of hard-wired neuronal circuits so that they adapt to the changing environment of an animal, and often alter global brain states such as the effect of norepinephrine on sleep and wakefulness. Neuromodulator molecules also play crucial roles in emotions, and are often released diffusely in a paracrine fashion. The effects of a neuromodulator on neurons are not confined to either excitatory or inhibitory; are diverse depending on the receptors; and are often relatively long-lasting, because most receptors are GPCRs.

Dopamine, norepinephrine, serotonin and histamine are amine neuromodulators. Dopamine and norepinephrine are classified as catechol amines. Dopamine is produced in neurons of the ventral tegmental area and the substantia nigra pars compacta, etc. Dopamine receptors are classified as D_1-type receptors mainly coupled to Gs protein and D_2-type receptors coupled to Gi protein. Dopamine neurons in the substantia nigra pars compacta send outputs to the dorsal striatum, and play roles in motor control through the direct and indirect pathways. There are two types of GABAergic inhibitory projection neurons in the dorsal striatum, those expressing D_1 receptor constituting the direct pathway and those expressing D_2 receptor constituting the indirect pathway. Dopaminergic neurons in the ventral tegmental area send axons to the nucleus accumbens, etc., and play roles in control of motivation, reward seeking, reinforcement learning, etc. The ventral tegmental area is implicated in positive reward and plays a crucial role in wanting of pleasure. In reinforcement learning dopamine neurons in the ventral tegmental area code a reward prediction error and enhance the occurrence of synaptic plasticity. The malfunction of dopaminergic systems has been related to various neurological and neuropsychiatric diseases.

Norepinephrine is converted from dopamine in neurons of the locus coeruleus. Norepinephrine is also called noradrenaline, and its receptor is usually called the adrenergic receptor. Adrenergic receptors are classified into 3 subtypes, α_1-, α_2- and β-adrenergic receptors. Norepinephrine is implicated in the control of arousal, response to stress, attention, learning, etc. The optogenetic activation of the locus coeruleus neurons promotes sleep to wake transition. In the sensory cortex, the overall effects of norepinephrine are decrease in spontaneous neuronal activities and increase in neuronal activities driven by a salient input, which seems to improve the signal to noise ratio of input signals. Norepinephrine in the prefrontal cortex is considered to play roles in top-down attention. The malfunction of norepinephrine system has been considered to be involved in various psychiatric disorders.

Serotonin or 5-hydroxytryptamine (5-HT) is indole amine. Serotonin has been considered to be involved in the control of sleep cycles, anxiety, nociception, etc. The malfunction of the serotonin system has been implicated in depression, anxiety disorders, etc. Selective serotonin reuptake inhibitors have been widely used as antidepressants. Histamine is a type of imidazole amine, and histaminergic neurons are located in the tuberomamillary nucleus in the posterior hypothalamus. Histamine is implicated in sleep-wake regulation, immunity, etc. Histaminergic neurons are active only during waking, maintaining wakefulness and attention. Acetylcholine which is an important neurotransmitter working at neuromuscular junctions, also plays roles as a neuromodulator in the brain. Acetylcholine containing neurons are found in the tegmental nuclei, the basal forebrain, etc., and send outputs to wide areas in the brain. Cholinergic systems have been implicated in arousal, attention, self-organization of cerebral cortical circuits, etc.

Enkephalins, dynorphins, endorphins and oxytocin are peptide neuromodulators. Enkephalins, dynorphins, endorphins were found as substances showing opioids-like effects. The microinjection of an opioid into restricted hot spots in the nucleus accumbens enhances hedonic reaction in rats. These results together with other findings support roles of opioids in pain suppression and liking of pleasure. Oxytocin is a peptide hormone synthesized in neurons of some nuclei in the hypothalamus. Oxytocin neurons project to reward-related and fear-related regions such as the nucleus accumbens, the ventral tegmental area, etc. Oxytocin shows anxiolytic and antistress effects, and contributes to increase sociality.

The search for endogenous substances binding to cannabis receptors found anandamide and 2-arachidonoylglycerol (2-AG), which are called endocannabinoids. Anandamide and 2-AG are synthesized from phospholipids on the cell membrane when neurons are activated. Endocannabinoid produced in a postsynaptic neuron, travels retrogradely to a presynaptic terminal and binds to CB1 receptor, resulting in suppression of neurotransmitter release.

Opioids, cannabinoids and their related substances are subject to abuse and may cause addiction. Opioids such as morphine and heroin induce euphoria and sedation, that is a drowsy state with reduced arousal. Thus, they are classified to depressants. Opioids seem to induce euphoria through their actions in the nucleus accumbens. In addition, opioids increase the extracellular dopamine level in the nucleus accumbens, which might cause wanting of pleasure and addiction. Cocaine, methamphetamine, etc. induce euphoria, increase alertness and reduce fatigue. Thus, they are classified as stimulants. Cocaine, etc. increase the concentration of dopamine and norepinephrine in the extracellular space, and seem to cause wanting and alerting. Cannabis such as marijuana affects the endocannabinoid system and causes euphoria, relaxation, etc. They reduce GABA release causing disinhibition of dopaminergic neurons and increase the dopamine level. Thus, dopamine signaling seems to be critically involved in addiction.

Neuronal activities in certain brain areas can be altered not only by neuromodulator molecules but also by various physical stimulation methods. Optogenetic stimulations, electrical stimulations, magnetic stimulations and ultrasound stimulations of some brain regions have been used for research and therapeutic purposes.

References

Anghel, D.C., G.V. Nitescu, A. Tiron, C.M. Gutu and D.I. Baconi. 2023. Understanding the mechanisms of action and effects of drugs of abuse. Mol. 28: 4969.

Alberts, B., K. Hopkin, A. Johnson, D. Morgan, K. Roberts, P. Walter et al. 2023. Essential Cell Biology. 6th Ed. Norton, New York, NY.

Belelli, D., G.D. Phillips, J.R. Atack and J.J. Lambert. 2021. Relating neurosteroid modulation of inhibitory neurotransmission to behaviour. J. Neuroendocrinol. 34: e13045.

Berger, M., J.A. Gray and B.L. Roth. 2009. The expanded biology of serotonin. Annu. Rev. Med. 60: 355–366.

Benarroch, E.E. 2012. Periaqueductal gray: an interface for behavioral control. Neurol. 78: 210–217.

Bromberg-Martin, E.S., M. Matsumoto and O. Hikosaka. 2010. Dopamine in motivational control: rewarding, aversive, and alerting. Neuron 68: 815–834.

Bucher, D. and E. Marder. 2013. Snapshot: Neuromodulation. Cell 155: 482.

Carter, M.E., O. Yizhar, S. Chikahisa, H. Nguyen, A. Adamantidis, S. Nishino et al. 2010. Tuning arousal with optogenetic modulation of locus coeruleus neurons. Nat. Neurosci. 13: 1526–1533.

Che, T. and B.L. Roth. 2023. Molecular basis of opioid receptor signaling. Cell 186: 5203–5219.

Cowan, P.J. 2008. Serotonin and depression: pathophysiological mechanism or marketing myth? Trends Pharmacol. Sci. 29: 433–436.

Cristino, L., T. Bisogno and V. Di Marzo. 2020. Cannabinoids and the expanded endocannabinoid system in neurological disorders. Na. Rev. Neurol. 16: 9–29.

Darmani, G., T.O. Bergmann, K.B. Pauly, C.F. Caskey, L. de Lecea, A. Fomenko et al. 2022. Non-invasive transcranial ultrasound stimulation for neuromodulation. Clinic. Neurophysiol. 135: 51–73.

Floresco, S.B., A.R. West, B. Ash, H. Moore and A.A. Grace. 2003. Afferent modulation of dopamine neuron firing differentially regulates tonic and phasic dopamine transmission. Nat. Neurosci. 6: 968–973.

Fricker, L.D., E.B. Margolis, I. Gomes and L.A. Devi. 2020. Five decades of research on opioid peptides: current knowledge and unanswered questions. Mol. Pharmacol. 98: 96–108.

Grinevich, V. and I.D. Neumann. 2021. Brain oxytocin: how puzzle stones from animal studies translate into psychiatry. Mol. Psychiat. 26: 265–279.

Haas, H., O.A. Sergeeva and O. Selbach. 2008. Histamine in the nervous system. Physiol. Rev. 88: 1183–1241.

Hariz, M. and P. Blomstedt. 2022. Deep brain stimulation for Parkinson's disease. J. Intern. Med. 292: 764–778.

Henderson, J., T. Federici and N. Boulis. 2009. Optogenetic neuromodulation. Neurosurgery 64: 796–804.

Hensler, J.G. 2006. Serotonergic modulation of the limbic system. Neurosci. Biobehav. Rev. 30: 203–214.

Hirano, T and T. Inoshita. 2021. Contribution of norepinephrine to cerebellar long-term depression and motor learning. pp. 337–348. *In*: Mizusawa, H. and S. Kakei. (eds.). Cerebellum as a CNS Hub. Springer, New York, NY.

Joëls, M. 2018. Corticosteroids and the brain. J. Endocrinol. 238: R121–R130.

Klein, M.O., D.S. Battagello, A.R. Cardoso, D.N. Hauser, J.C. Bittencourt and R.G. Correa. 2019. Dopamine: functions, signaling, and association with neurological diseases. Cell. Mol. Neurobiol. 39: 31–59.

Knobloch, H.S., A. Charlet, L.C. Hoffmann, M. Eliava, S. Khrulev, A.H. Cetin et al. 2012. Evoked axonal oxytocin release in the central amygdala attenuates fear response. Neuron 73: 553–566.

Knotkova, H., C. Hamani, E. Sivanesan M.F.E. Le Beuffe, J.Y. Moon, S.P Cohen et al. 2021. Neuromodulation for chronic pain. Lancet 397: 2111–2124.

Krauss, J.K., N. Lipsman, T. Aziz, A. Boutet, P. Brown, J.W. Chang et al. 2021. Technology of deep brain stimulation: current status and future directions. Nat. Rev. Neurol. 17: 75–87.

Lee, S. and Y. Dan. 2012. Neuromodulation of brain states. Neuron 76: 209–222.

Lerner, T.N., A.L. Holloway and J.L. Seiler. 2021. Dopamine, updated: reward prediction error and beyond. Curr. Opin. Neurobiol. 67: 123–130.

Lu, H. and K. Mackie. 2021. Review of the endocannabinoid system. Biol. Psychiat. 6: 607–615.

Luo, L. 2021. Principles of Neurobiology. 2nd Ed. CRC Press, Boca Raton, FL.

Marsh, N., A.A. Marsh, M.R. Lee and R. Hurlemann. 2021. Oxytocin and the neurobiology of prosocial behavior. Neurosci. 27: 604–619.

McCall, J.G., R. Al-Hasani, E.R. Siuda, D.Y. Hong, A.J. Norris, C.P. Ford et al. 2015. CRH engagement of the locus coeruleus noradrenergic system mediates stress-induced anxiety. Neuron 87: 605–620.

Mohammad-Zadeh, L.F., L. Moses and S.M. Gwaltney-Brant. 2008. Serotonin: a review. J. vet. Pharmacol. Therap. 31: 187–199.

Nichols, D.E. and C.D. Nichols. 2008. Serotonin receptors. Chem. Rev. 108: 1614–1641.

Pecina, S. and K.C. Berridge. 2005. Hedonic hot spot in nucleus accumbens shell: where do mu-opioids cause increased hedonic impact of sweetness? J. Neurosci. 25: 11777–11786.

Picciotto, M.R., M.J. Higley and Y.S. Mineur. 2012. Acetylcholine as a neuromodulator: cholinergic signaling shapes nervous system function and behavior. Neuron 76: 116–129.

Picón-Pagès, P., J. Garcia-Buendia and F.J. Muñoz. 2018. Functions and dysfunctions of nitric oxide in brain. Biochim. Biophys. Acta. Mol. Basis Dis. 1865: 1949–1967.

Purves, D., G.J. Augustine, D. Fitzpatrick, W.C. Hall, A.S. LaMantia, R.D. Mooney et al. 2019. Neuroscience. 6th Ed. Sinauer, Sunderland, MA.

Remage-Healey, L. 2014. Frank Beach award winner: steroids as neuromodulators of brain circuits and behavior. Horm. Behav. 66: 552–560.

Rudolph, L.M., C.A. Cornil, M.A. Mittelman-Smith, J.R. Rainville, L. Remage-Healey, K. Sinchak et al. 2016. Actions of steroids: new neurotransmitters. J. Neurosci. 36: 11449–11458.

Saigusa, T., Y. Aono and J.L. Waddington. 2017. Mechanisms underlying δ- and μ-opioid receptor agonist induced increases in extracellular dopamine level in the nucleus accumbens of freely moving rats. J. Oral Sci. 59: 195–200.

Sara, S.J. 2009. The locus coeruleus and noradrenergic modulation of cognition. Nat. Rev. Neurosci. 10: 211–223.

Scammell, T.E., A.C. Jackson, N.P. Franks, W. Wisden and Y. Dauvilliers. 2019. Histamine: neural circuits and new medications. Sleep 42: zsy183.

Schultz, W., P. Dayan and P.R. Montague. 1997. A neural substrate of prediction and reward. Sci. 275: 1593–1599.

Schwarz, L.A. and L. Luo. 2015. Organization of the locus coeruleus-norepinephrine system. Curr. Biol. 25: R1051–R1056.

Shastry, S., J. Hu, M. Ying and X. Mao. 2023. Cell therapy for Parkinson's disease. Pharmaceutics 15: 2656.

Speranza, L., U. di Porzio, D. Viggiano, A. de Donato and F. Volpicelli. 2021. Dopamine: the neuromodulator of long-term synaptic plasticity, reward and movement control. Cells 10: 735.

Terenius, L. and B. Johansson. 2010. The opioid systems – panacea and nemesis. Biochem. Biophy. Res. Commun. 396: 140–142.

Thiele, A. and M.A. Bellgrove. 2018. Neuromodulation of attention. Neuron 97: 769–785.

Tully, K. and V. Bolshakov. 2010. Emotional enhancement of memory: how norepinephrine enables synaptic plasticity. Mol Brain 3: 15.

van den Pol, A.N. 2012. Neuropeptide transmission in brain circuits. Neuron 76: 98–115.

10

Sleep and Dream

Evolutional Origin and Purposes of Sleep

Sleep can be defined as a naturally recurring reversible state of perceptual disengagement, reduced consciousness and relative immobility (Grandner and Fernandez 2021). When we fall asleep, we lose consciousness and the universe vanishes for us (Tononi et al. 2024). Sleeping animals do not eat or mate, and are vulnerable targets of predators. Therefore, there are certain demerits but apparently no clear merit in sleeping for the survival or reproduction of animals. Nevertheless, most animals sleep or show sleep-like phenomena, and sleep deprivation causes deleterious effects at least in some animals including humans. Thus, sleep seems to be indispensable (Anafi et al. 2019).

Throughout the various animal species, sleep may be redefined as quiescent behavioral states associated with reduced responsiveness to weak stimuli and rapid reversibility in response to strong stimuli. In mammals sleep states can be also inferred by brain waves as explained in a following section. Not only vertebrates but also invertebrates, such as insects, show sleep-like phenomena. Cockroaches, honeybees and fruit fries show inactive or immobile periods with decreased responsiveness to external stimuli (Tobler 1983, Kaiser and Steiner-Kaiser 1983, Shaw et al. 2000). A nematode has periods during which it reduces movements and does not respond to weak stimuli (Raizen et al. 2008). A jerry fish having a dispersed nervous system without a brain or a central nervous system, also shows inactive periods. The frequency of pulsations is higher during the day as compared to those during the night, and responsiveness to stimulus is reduced at night (Nath et al. 2017). Cetaceans such as dolphins and whales are supposed to be unable to have completely inactive states because of their need to breathe in the sea. Sometimes, they show characteristic brain waves representing sleep states in one side of the brain, while the other side of the brain seems to be awake for a certain period (Anafi et al. 2019). Thus, sleep or sleep-like phenomena are prevalent, suggesting their importance in our lives.

Studies on molecular changes associated with sleep and sleep-like phenomena in different tissues of various animals suggest that metabolic demands are a key driver of sleep, and that primarily sleep plays metabolic roles (Anafi et al. 2019). Conservation

DOI: 10.1201/9781003518822-10

of energy during the time when an animal has difficulty to get food such as at night; restoration of key molecules that are consumed during waking periods; and clearance of metabolic waste products, have been proposed as functional roles of sleep. In addition, an idea that sleep supports learning and memory has also been proposed. Sleep has been suggested to be a brain state optimizing memory consolidation, whereas an awake state has been suggested to be optimized for encoding memories (Rasch and Born 2013). During slow wave sleep, that is explained in a later section, neuronal activities which seem to reflect memory replay occur in the hippocampus. It has been proposed that the hippocampal replay contributes to the transformation of hippocampus-dependent episodic memory to the neocortical memory (Brodt et al. 2023).

Necessity of Sleep and Variation of Sleep Time

Continued awake periods for several days in a human causes involuntary sleep even if the human keeps trying to stay awake. Sleep disruption causes functional impairment in emotion regulation, recognition, memory, etc. in general, and is also related to mental disorders such as depression, anxiety disorder, etc. Sleep duration outside of the normative seven to eight hours range in humans has also been related to cardiovascular diseases, diabetes, disfunction in the immune system such as increased rates of infection and blunted vaccine responses (Grandner and Fernandez 2021).

A sleep disruption experiment using rats was performed. A pair of rats were housed in an experimental setup, and their brain waves were recorded as electroencephalogram, which is explained further in a later section. Whenever an experimental rat was about to fall asleep, which was detected by the electroencephalogram, awakening stimulation was applied to both rats. The stimulation completely disrupted the sleep of the experimental rat, while the control rat could sleep while the electroencephalogram of the experimental rat showed an aroused pattern (Rechtschaffen et al. 1983). The experimental rat died within a few weeks of sleep disturbance, indicating the necessity of sleep for staying alive. The cause of death was unclear, as the experimental rat showed multiple abnormalities such as weight loss, compromised thermal regulation, weakening of the immune system, loss of gut integrity, etc. (Luo 2021).

Adequate duration of sleep varies depending on the age and is different for different species. The recommended duration of sleep for adult humans is at least seven hours per night, and some researchers advise that adults should not aim to have more than nine hours of sleep. The recommended duration of daily sleep for children is longer than that for adults: infants require the longest sleep of around seventeen hours, and toddlers a shorter sleep of nine hours (Purves et al. 2019, Grandner and Fernandez 2021). Among mammals, sleep duration varies profoundly ranging from a few hours in large herbivores to twenty hours in brown bats. Certain animal species such as cetaceans and birds can greatly curtail sleep time during particular developmental periods or seasons. During migration, a great frigatebird *Fregata* spends less than one-tenth of the time sleeping when flying over the ocean, as compared with the time sleeping on land. A newborn killer whale *Orcinus orca* is thought to be almost continuously active in the first month after the birth (Anafi et al. 2019). Thus, the duration of necessary sleep time is variable among animal species and situations.

Circadian Rhythm

In most animals, sleep timing is determined by the twenty-four hours activity cycle, called the circadian rhythm, based on the daily cycle of light and darkness caused by the rotation of the earth. However, the circadian rhythm can be retained without change in light. Mice kept in continuous darkness show rhythmic activity change with a cycle that is slightly shorter than twenty-four hours, implying that they have an internal clock. The application of light resets the cycle, and is called light entrainment (Luo 2021).

Studies on fruit flies and mice have identified molecules contributing to the formation of the circadian rhythm such as Period, Clock, CYC, BMAL1, Timeless and CRY. Among them, CYC and Timeless are found in fruit fries, whereas BMAL1 and CRY were found in mice. Their interactions and transcription regulations generate the circadian rhythm, which occur not only in neurons but also in almost all peripheral cells (Patke et al. 2020, Luo 2021).

In mammals the suprachiasmatic nuclei, located above the optic chiasma in the hypothalamus (Figure 6 of Chapter 4), are the center in both the regulation of the circadian rhythm and the light entrainment, and the latter is enabled by outputs from intrinsically photosensitive retinal ganglion neurons (Figure 4 of Chapter 4) which express melanopsin as a photosensor molecule. Activities of neurons in the suprachiasmatic nuclei are well synchronized and oscillate with the circadian rhythm in isolation. Neurons in the suprachiasmatic nuclei are mostly GABAergic and show high activities in the light period. These neurons send outputs to many nuclei in the hypothalamus which regulate various peripheral organs and also wake- and sleep-promoting systems in the brain, that are explained in a following section (Purves et al. 2019, Scammell et al. 2017, Luo 2021). The hypothalamus has been explained in chapter eight.

Levels of Sleep and Brain Waves (Electroencephalogram)

Sleep is not a single state, but there are several states which can be distinguished by brain waves or the electroencephalogram (EEG), which reflects the collective activities of many neurons in the brain and is recorded by electrodes placed on the surface of the scalp. In an awake state an EEG shows high frequency (β frequency of 15–60 Hz) and low amplitude (around 30 mV) oscillations. When an animal or a human falls asleep, low frequency oscillations increase in amplitude. Three or four sleep stages I–III or I–IV are distinguishable. In addition, a different type of sleep, named rapid eye movements (REM) sleep is known, which is explained further in the following section (Purves et al. 2012, Luo 2021).

Sleep states, other than REM sleep, corresponding to stage I–III or I–IV, are called non-REM sleep. Within non-REM sleep, stage I is a brief falling asleep stage showing EEG with θ (4–7 Hz) frequency and 50–100 mV amplitude oscillations,

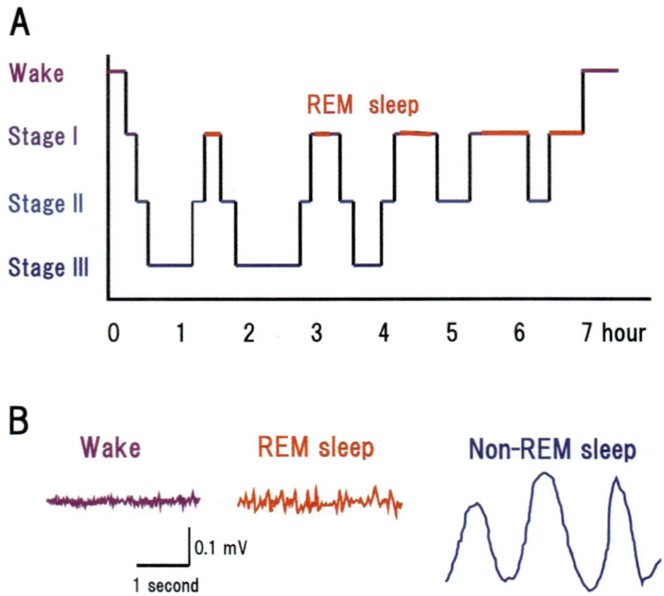

C

	Wake	REM sleep	Non-REM sleep
Ach	High	High	Low
Amines	High	Low	Low

Figure 1. Schematic presentation of sleep states, EEGs and activation levels of acetylcholine and amine systems. A, Changes of sleep states during a night. Sleep stage I-III are non-REM sleeps, and stage III is the deepest. From stage I non-REM sleep transition to REM sleep (red lines) takes place. Changes of sleep stages occur cyclically. B, Schematic drawings of EEGs during an awaked state, REM sleep and non-REM stage III sleep. Frequencies and amplitudes of EMGs are different among these states. C, Activity levels of acetylcholine and amine systems during an awaked state, REM sleep and non-REM sleep in the brain (Scammell et al. 2017, Purves et al. 2019, Luo 2021).

stage II is a light sleep showing EEG with slow oscillations and occasional spindles that are α (10–12 Hz) frequency and 50–150 mV amplitude activities. The interactions between the thalamic and cerebral cortical neurons cause spindles (Purves et al. 2019). Stage III is a deep sleep stage showing EEG with slower δ (0.5–4 Hz) frequency and large amplitude (100–150 mV) waves. Sleep starts from a light stage I non-REM sleep, and descends to deep stage II and III sleeps. Then sleep ascends to lighter stages, and at the light stage transition to REM sleep takes place. After that, the transition to non-REM sleep and descending to deeper sleeps follow. Such sleep cycles occur several times before we wake up (Figures 1A, B). Stronger sensory stimuli are required to wake up a person from a deeper sleep stage (Purves et al. 2019, Scammell et al. 2017, Luo 2021, Tononi et al. 2024).

The slow and large amplitude δ waves in deep sleep stage III are brought about by the synchronized electrical activities of cerebral cortical neurons showing up and down states (Jercog et al. 2017), which partly depend on activities of the thalamic neurons projecting to the cortex. These thalamocortical neurons show two states of activities; a bursting or an oscillatory state, and a tonically active state, as explained in chapter eight. In stage III sleep, the thalamocortical neurons are hyperpolarized and show synchronized oscillating activities, and the transmission of sensory information to the cerebral cortex is suppressed. On the other hand, in an awake state the thalamocortical neurons are tonically active and send sensory information, which comes from the peripheries (Steriade et al. 1993, Purves et al. 2019).

REM Sleep

One conspicuous characteristic of REM sleep is the rapid and ballistic movements of eyeballs, from which the name REM sleep has been derived, although REM sleep has other characteristics such as an EEG pattern that is somewhat similar to an awake state (Figure 1B), prevalence of dreaming, penile erection, paralysis of many large skeletal muscles, etc. Some humans show sleep walk, which you might think takes place during REM sleep. But it does not, because in REM sleep large skeletal muscles are paralyzed. During non-REM sleep, decreases in the blood pressure, the metabolic rate, the heart rate, etc. occur, whereas in REM sleep they are increased. REM sleep takes place in all mammals, and at least in some birds and some reptiles. As explained above REM sleep is somewhat similar to an awake state, but there is a clear difference, which is that the brain is disconnected from sensory signals and neuronal activities are solely driven internally during REM sleep. The concentrations of amine neuromodulators are different between awake and REM sleep states as explained further in the following section. REM sleeps take about 20% of the whole sleep time (Purves et al. 2019, Luo 2021, Tononi et al. 2024).

Neuronal Mechanisms of Sleep Regulation

There are arousal systems which cause or maintain an awakened state in the brain. One is the cholinergic or acetylcholine system. Cholinergic neurons in the tegmental nuclei send outputs to the thalamus and to cholinergic neurons in the basal forebrain, which in turn project to many regions of the forebrain. These cholinergic neurons are active during wakefulness and REM-sleep, and influence activities of the thalamic and cerebral cortical neurons, and work to arouse an animal or a human (Figure 1C) (Lee and Dan 2012, Luo 2021).

Another arousal system is aminergic and consists of norepinephrine neurons in the locus coeruleus, serotonin neurons in the raphe nuclei, histamine neurons in the tuberomammillary nucleus and dopamine neurons in the ventral periaqueductal grey (PG) (Figure 1 of Chapter 8). These aminergic neurons are active during wakefulness, and decrease activity not only during non-REM sleep, but also during REM sleep (Figure 1C), except dopamine neurons which are implicated in initiation of REM sleep (Hasegawa et al. 2022). Thus, activities of aminergic neurons are clearly different between in an awaked state and in REM sleep. Aminergic neurons also contribute to the maintenance of an aroused state. Thus, aminergic neuromodulator molecules together with acetylcholine play critical roles in the regulation of awake and sleep states (Lee and Dan 2012, Luo 2021).

In addition to cholinergic and aminergic systems, there is another arousal system using peptide orexin or hypocretin (Lin et al. 1999). Neurons expressing orexin are localized in the lateral hypothalamus, and send outputs to wide brain areas, including cholinergic and aminergic neurons, as described above. Orexin neurons are active during wakefulness and become silent during sleep, and the optogenetic activation of orexin neurons promotes waking up from sleep. Orexin neurons have been considered to support wakefulness through the activation of cholinergic and aminergic systems. The dysfunction of the orexin system is a cause of narcolepsy, which is a kind of sleep disorder showing difficulty to stay awake when exited in particular (Lin et al. 1999, Luo 2021).

There are also neuronal systems that promote sleep. One is GABAergic neurons in the preoptic area of the hypothalamus. These neurons are active during sleep and send outputs to cholinergic and aminergic neurons as explained above, and inhibit their activities. On the other hand, these GABAergic neurons are innervated by neurons in the cholinergic and aminergic arousal systems, and the activities are suppressed by acetylcholine, norepinephrine, dopamine and serotonin. Thus, arousal systems and sleep-promoting systems inhibit each other. The optogenetic activation of the GABAergic neurons in the preoptic area projecting to the tuberomammillary nucleus promotes sleep (Chung et al. 2017).

Neurons in the suprachiasmatic nuclei, which are at the center of the circadian rhythm regulation, send outputs to other nuclei in the hypothalamus including the subparaventricular zone, which in turn influences activities of wake-and sleep-promoting systems described above, mainly through the dorsomedial nucleus of the hypothalamus (Scammell et al. 2017).

Different Types of Dreams and Their Roles

Dreaming is prevalent during REM sleep, although it also occurs during non-REM sleep. Dreams during REM sleep and those during non-REM sleep have different characteristics. The formers tend to be long, vivid, complicated and story-like, whereas those during non-REM sleep are simple and short such as just a single scene or thoughts. One sometimes thinks that a dream in non-REM sleep was not a dream and that the dreamer was awake. It is also to be noted that the contents of dreams during REM sleep are often bizarre or weird, and distorted in time and space. The above information about dreams has been obtained by waking up research participants from REM or non-REM sleep. Otherwise, people often do not remember the contents of their dreams when they wake up from sleep (Siclari et al. 2017, Tsunematsu 2023, Tononi et al. 2024). One-third of the non-REM sleep awakenings yield null report that a research participant has been non-being or unconscious, another one-third yield white dreams whose contents cannot be remembered, and the rest one-third yield some remembered contents. On the other hand, most of REM sleep awakenings yield dream reports, although there are occasional reports of white dreams (Tononi et al. 2024).

In almost all dreams in REM sleep, we do not recognize that we are dreaming and accept the narrative as it is without questioning. However, some people have dreams in which they know that they are dreaming. Such dreams are called lucid dreams. In a lucid dream, cognitive control such as the ability to steer the dream narrative in a desired direction is possible (Baird et al. 2019). I, myself, do not have lucid dreams except for a few dreams. My first lucid dream which I remember, took place several days after I had learned about lucid dreams. In the dream I was walking on a stone-paved slope, and I noticed I was in a dream. Then, I tried to see the cobblestones of the slope, and could see details of cobblestones. After that experience, I noticed that I was dreaming several times, and tried to control the dream so that it became a pleasant one, but I have failed so far.

Roles or functional significances of dreams and whether animals other than humans have dreams, have been explored and/or speculated, but they remain enigmatic (Manger and Siegel 2020, Tsunematsu 2023, Tononi et al. 2024). The proposed functional roles of dreams are memory consolidation, elimination of unwanted memories, contribution to creativity, a rehearsal of threat perception and threat avoidance etc. Some researchers consider that dreams do not have any functional roles and are byproducts of random processes in the brain during REM sleep (Tsunematsu 2023, Tononi et al. 2024). It has been pointed out that some mammals do not show typical REM sleeps and they are not likely to have a type of dream which occurs during REM sleeps in humans, and that some aquatic mammals such as cetaceans showing unihemispheric sleep might not experience vivid dreams (Manger and Siegel 2020).

Neural Correlates of Dreams

fMRI recording during the sleep-onset periods in humans demonstrated that the visual experience during stage I or II non-REM sleep which was reported verbally, was represented by cerebral cortical activity patterns which were shared with visual stimulus perceptions (Horikawa et al. 2013). In another study, high-density electroencephalogram was recorded from human research participants during REM and non-REM sleeps. The participants were serially awakened during the night and asked about the dreams that they had just before awakening. The reports of participants of the dream they experienced were associated with the local decrease in low-frequency activities and the increase in high-frequency activities of the posterior cerebral cortical regions, involved in the processing of the sensory information in both REM and non-REM sleeps. In REM sleeps the regions showing increased high-frequency activities correlated with dream contents such as faces, speech, etc. (Siclari et al. 2017). These studies indicate that dream contents are based on neuronal activities in the brain, the cerebral cortices in particular, which are involved in processing a particular type of sensory information. Neuronal activities in a REM-sleep dream are discussed again in chapter thirteen.

The distinction between sleep and wakefulness might not be so strict. Mind wondering and/or drowsy awake states might be contiguous to stage I non-REM sleep. Neuronal activities in various brain regions during awake, non-REM and REM sleep states can be largely heterogeneous and asynchronous, and it seems possible that some regions show characteristics of non-REM sleep, while other regions show characteristics of awake state, which may be somewhat similar to the state of dolphin's brain during unihemispheric sleep (Tononi et al. 2024).

Self and Consciousness in Dreams

Dreams are purely personal and subjective experiences, just like consciousness in an awake state, and we can learn about another person's dreams only through verbal reports. As described above most reports tell us that dreams in REM sleeps are bizarre and distorted in time and space. In such dreams we usually do not pause to assess our experiences and just go on dreaming passively, accepting whatever thoughts or images that come to the mind without questioning. It is also said that we do not realize that we are dreaming, do not control our thoughts and do not imagine ourselves in the past or in the future. It has also been said that we are also impaired in the ability to analyze situations intelligently, to question assumptions critically, to reason properly, and to make appropriate decisions. Holding contradictory beliefs also takes place, and we easily accept impossible events or situations in dreams (Tononi et al. 2024). Exceptions are lucid dreams as explained in a previous section.

My dream experiences have also been very bizarre and distorted in time and space. In addition, sudden changes in my age, places, situations, accompanied persons, etc. occurred. In most cases, I accepted every image, thoughts, etc. in a dream, and my behavioral choices in dreams were sometimes very odd. They were choices that I would never make in an awakened state. Thus, self in dreams and self in an awake state seem quite different, at least in my case. However, as far as I remember, in some of my dreams I have questioned situations, felt strange and wondered whether I might be dreaming. As I have stated above, I have had a few lucid dreams, and I feel or think that awaken states, lucid dreams and other dreams, might not be so distinct and could be somewhat continuous.

Now, let me explain some of my dream experiences. In the first dream, I got on a train from a station with a few colleagues. In the next moment, I was alone on a bus, and then suddenly, I was walking down a road, when I noticed that I did not know where I was, and became anxious about whether I would be in time to get on an airplane: and then I awakened.

In the second dream, I was giving a lecture in the university, and in the next moment, I became a high school student taking a class. At that moment, I wondered whether I was a high school student or not, just for a while, then became anxious regarding whether I had been attending classes sufficiently and whether I could pass the entrance examination of a university, and awakened.

In the third dream, I was watching a battle of Japanese soldiers in middle ages as if I was watching a movie. I saw a commander of soldiers who was attending a meeting of commanders, then it seems that I became the commander. After a while, enemies suddenly attacked us and I was astonished and woke up. I am not quite sure whether I really became a commander or I experienced affairs through a commander's view.

The last dream I would like to explain, was the one that I had a few hours before and remember well. At first, I was seeing a couple, a young man and a woman, then the man became myself and the woman became my girl-friend. I walked with her for a while, then she suddenly became a cat. After a while two snakes appeared, and the cat started fighting with a larger snake. I was watching the fight worrying that the cat might be injured. Then, the cat left the fight and we went to a meeting in which a discussion about our marriage took place, then the cat or my girlfriend disappeared and I wondered where my girlfriend or the cat went. The meeting changed to another one in which a famous economist who was my father's friend gave a speech. Then, the meeting changed to a party, in which many different kinds of cakes and some hot dishes were prepared. I tried to have a dish but failed. Then, I left the room and went to the entrance hall, where I found a river and there was nobody there, and I woke up. In this dream, I did not recognize it was a dream, and also did not feel strange about the sudden change of a woman into a cat, etc.

Now, please allow me to consider and speculate about my dreams. All the scenes were bizarre. The sudden changes of situations and people, and other impossible events such as a change of a woman into a cat took place. I did not question these strange situations in the dreams, and accepted them as they were, although I thought of future plans within the dream context in some cases. Although the sudden changes of situations occurred often in my dreams, I notice that there was some continuation or relation between the situations. For example, in the first dream I was travelling throughout the dream, although the vehicles changed abruptly. All the contents of the second dream were related to the university or school, although my status and age suddenly changed.

I speculate that the continuation of contents in a dream might be brought about by waves of neuronal activities travelling from some neuronal assemblies to other neuronal assemblies representing related objects, just as it occurs during the thinking processes in an awakened state. I also wonder if sudden changes in situations in dreams could be ascribed to decreased activities of the prefrontal cerebral cortices and/or weakened attentional maintenance. These might be based on the decreased activities in the norepinephrine system, which might make it difficult to keep neuronal activities related to a particular object at a high level. The propagation of waves of neuronal activities, which does not usually occur in an awakened state, might occur in dreams, which might be the reason for sudden changes of situations, etc. Such travelling of neuronal activity waves during a dream, might also take place between neuronal assemblies representing objects which are considered unrelated, and could contribute to finding novel relationships between these objects, affairs and/or notions in some cases. Such a finding could be regarded as highly creative in a few cases, as I think that discovery, invention, originality and/or creativity are often based on novel findings of previously unrecognized relationships. I also note that anxiety or fear often aroused me in dreams, which might be a reason why contents of most dreams we remember are negative. Such arousal by fear or anxiety might be caused by the activation of norepinephrine systems, which are related to both fear, anxiety and arousal as explained in chapters eight and nine.

Now I want to compare self in an awakened state and in REM sleep. I cannot be responsible for my thoughts, decisions and behavioral selections made by me in my dreams. The self in dreams seems to have different characteristics, feelings and personality, although both selves in an awakened state and in a dream, should depend on the same brain. The differences in working states of brain areas or neuronal systems such as the frontal cortex, the attention systems and the aminergic systems, seem to contribute to different characteristics of self in an awakened state and in a dream. Very different global functional states of the brain seem to exist, and the alteration seems to depend on the neuromodulator systems. The center of self or of metacognitive processing, seems to be different among different states such as an awaked state, REM sleep and non-REM sleep. I also wonder if the center might not be confined to one even in awake states, about which I discuss again in chapter thirteen.

Summary and Conclusion

Sleep can be defined as a naturally recurring reversible state of perceptual disengagement, reduced consciousness and relative immobility. Sleeping animals do not eat or mate and are vulnerable targets of a predator, which are clear demerits. However, most animals sleep or show sleep-like phenomena, and sleep deprivation causes deleterious effects at least in some animals. Metabolic demands have been considered a key driver of sleep. Conservation of energy, restoration of key molecules and clearance of metabolic waste products, have been proposed as functional roles of sleep. Sleep has been also suggested to be involved in memory consolidation. Sleep usually shows daily cycles called the circadian rhythm. The suprachiasmatic nuclei in the hypothalamus are the center of the regulation of the circadian rhythm. Neurons in the suprachiasmatic nuclei show high activities in the light period, and send outputs to many nuclei in the hypothalamus through which wake- and sleep-promoting neuronal systems are influenced.

There are several sleep states which can be distinguished by the electroencephalogram (EEG), which reflects the collective activities of many neurons in the brain. Three sleep stages I–III in non-REM sleep and a different type of sleep named rapid eye movements (REM) sleep are distinguished. Sleep starts from a light stage I sleep and descend to deeper stage II and III. Then, sleep ascends to lighter stages, and at a light stage transition to REM sleep occurs. After that, transition to non-REM sleep and descending to deeper sleep stages follow. Such cycles occur several times during sleep. Stronger sensory stimuli are required to wake up a person from a deep sleep stage. The slow and large amplitude EEG in deep sleep stage III is brought about by synchronized electrical activities of cerebral cortical neurons. REM sleep is characterized by rapid and ballistic movements of eyeballs, EEG pattern somewhat similar to an awakened state, prevalence of dreaming, etc. During non-REM sleep decreases in the blood pressure, the metabolic rate, the heart rate, etc. occur, whereas in REM sleep they are increased.

There are arousal systems which cause or maintain an awakened state. One is cholinergic or acetylcholine systems in tegmental nuclei and in the basal forebrain. The cholinergic neurons are active during wakefulness and REM-sleep, influence the activities of thalamic and cerebral cortical neurons, and work to arouse an animal. Other arousal systems are aminergic consisting of norepinephrine, serotonin, histamine and dopamine neurons. These aminergic neurons are active during wakefulness, and decrease activities not only during non-REM sleep, but also during REM sleep. I note that some dopamine neurons are implicated in the initiation of REM sleep. There is another arousal system using peptide orexin. Neurons expressing orexin are localized in the lateral hypothalamus, and send outputs to wide brain areas, including the cholinergic and aminergic neurons. There are also neuronal systems that promote sleep such as GABAergic neurons in the preoptic area of the hypothalamus. These neurons are active during sleep and send outputs to cholinergic and aminergic neurons explained above, and inhibit their activities. On the other hand, these GABAergic neurons are innervated by neurons in the cholinergic and aminergic arousal systems, and the activities are suppressed by acetylcholine, norepinephrine, dopamine and serotonin.

Dreaming is prevalent during REM sleep, although it also occurs during non-REM sleep. Dreams during REM sleep tend to be long, vivid, complicated and story-like, whereas those during non-REM sleep are simple and short, such as just a single scene or thought. Dreams during REM sleep are often bizarre, and distorted in time and space, and in it we do not recognize that we are dreaming and accept the narrative as it is. fMRI recording and high-density electroencephalogram recording from human research participants who were serially awakened during REM and/or non-REM sleeps, showed that dream contents are based on neuronal activities in the cerebral cortical regions, which are involved in processing a particular type of sensory information. The proposed roles of dreams are memory consolidation, elimination of unwanted memories, contribution to creativity, a rehearsal of threat perception, etc.

Although sudden changes of situations occur often in REM dreams, there is some continuation between the situations. These continuations might be brought about by waves of neuronal activities travelling from some neuronal assemblies to other neuronal assemblies, similar to those that occur during the thinking processes in an awakened state. The sudden changes of situations in dreams might be ascribed to the decreased activities of the prefrontal cerebral cortices and/or weakened attentional maintenance, which might be related to decreased activities in the norepinephrine system and might make it difficult to keep neuronal activities related to a particular object or affair at a high level. The self in dreams seems to have different characteristics, feelings and personality, although both selves in an awake state and in dreams should depend on the same brain. The differences in working states of the brain areas seem to contribute to the different characteristics of self in an awakened state and in dreams. The center of self or metacognitive processing seems to be different between the two states.

References

Anafi, R.C., M.S. Kayser and D.M. Raizen. 2019. Exploring phylogeny to find the function of sleep. Nar. Rev. Neurosci. 20: 109–116.

Baird, B., S.A. Mota-Rolim and M. Dresler. 2019. The cognitive neuroscience of lucid dreaming. Neurosci. Biobehav. Rev. 100: 305–323.

Brodt, S., M. Inostroza, N. Niethard and J. Born. 2023. Sleep-A brain-state serving systems memory consolidation. Neuron 111: 1050–1075.

Chung, S., F. Weber, P. Zhong, C.L. Tan, T.N. Nguyen, K.T. Beier et al. 2017. Identification of preoptic sleep neurons using retrograde labelling and gene profiling. Nat. 545: 477–481.

Hasegawa, E., A. Miyasaka, K. Sakurai, Y. Cherasse, Y. Li and T. Sakurai. 2022. Rapid eye movement sleep is initiated by basolateral amygdala dopamine signaling in mice. Sci. 375: 994–1000.

Grandner, M.A. and F. Fernandez. 2021. The translational neuroscience of sleep: a contextual framework. Sci. 374: 568–573.

Horikawa, T., M. Tamaki, Y. Miyawaki and Y. Kamitani. 2013. Neural decoding of visual imagery during sleep. Sci. 340: 639–642.

Kaiser, W. and J. Steiner-Kaiser. 1983. Neuronal correlates of sleep, wakefulness and arousal in a diurnal insect. Nat. 301: 707–709.

Jercog, D., A. Roxin, P. Barthó, A. Luczak, A. Compte and J. de la Rocha. 2017. UP-DOWN cortical dynamics reflect state transitions in a bistable network. Elife 6: e22425.

Lee, S. and Y. Dan. 2012. Neuromodulation of brain states. Neuron 76: 209–222.

Lin, L., J. Faraco, R. Li, H. Kadotani, W. Rogers, X. Lin et al. 1999. The sleep disorder canine narcolepsy is caused by a mutation in the hypocretin (orexin) receptor 2 gene. Cell. 98: 365–376.

Luo, L. 2021. Principles of Neurobiology. 2nd Ed. CRC Press, Boca Raton, FL.

Manger, P.R. and J.M. Siegel. 2020. Do all mammals dream? J. Comp. Neurol. 528: 3198–3204.

Nath, R.D., C.N. Bedbrook, M.J. Abram, T. Basinger, J.S. Boi, D.A. Prober et al. 2017. The jellyfish *Cassiopea* exhibits a sleep-like state. Curr. Biol. 27: 2984–2990.

Patke, A., M.W. Young and S. Axelrod. 2020. Molecular mechanisms and physiological importance of circadian rhythms. Nat. Rev. Mol. Cell Biol. 21: 67–84.

Purves, D., G.J. Augustine, D. Fitzpatrick, W.C. Hall, A.S. LaMantia, R.D. Mooney et al. 2019. Neuroscience. 6th Ed. Sinauer, Sunderland, MA.

Raizen, D.M., J.E. Zimmerman, M.H. Maycock, U.D. Ta, Y. You, M.V. Sundaram et al. 2008. Lethargus is a *Caenorhabditis elegans* sleep-like state. Nat. 451: 569–572.

Rasch, B. and J. Born. 2013. About sleep's role in memory. Physiol. Rev. 93: 681–766.

Rechtschaffen, A., M.A. Gilliland, B.M. Bergmann and J.B. Winter. 1983. Physiological correlates of prolonged sleep deprivation in rats. Sci. 221: 182–184.

Scammell, T.E., E. Arrigoni and J.O. Lipton. 2017. Neural circuitry of wakefulness and sleep. Neuron 93: 747–765.

Shaw, P.J., C. Cirelli, R.J. Greenspan and G. Tononi. 2000. Correlates of sleep and waking in *Drosophila melanogaster.* Sci. 287: 1834–1837.

Siclari, F., B. Baird, L. Perogamvros, G. Bernardi, J.J. LaRocque, B. Riedner et al. 2017. The neural correlates of dreaming. Nat. Neurosci. 20: 872–878.

Steriade, M., D.A. McCormick and T.J. Sejnowski. 1993. Thalamocortical oscillations in the sleeping and aroused brain. Sci. 262: 679–685.

Tobler, I. 1983. Effect of forced locomotion on the rest–activity cycle of the cockroach. Behav. Brain Res. 8: 351–360.

Tononi, G., M. Boly and C. Cirelli. 2024. Consciousness and sleep. Neuron 112: 1568–1594.

Tsunematsu, T. 2023. What are the neural mechanisms and physiological functions of dreams? Neurosci. Res. 189: 54–59.

11

Psychiatric and Neurodevelopmental Disorders

Brain Disorders

Most brain disorders have been classified into psychiatric, neurodevelopmental and neurodegenerative disorders. In neurodegenerative disorders increasing numbers of neurons become dis-functional or die, and the brain shows significant atrophy. Psychiatric disorders affect the adult patient's mind without clear structural changes of the brain. Neurodevelopmental disorders first appear in infancy or in early childhood. However, recent progresses in studies on brain disorders have blurred such categorization. For example, some psychiatric disorders such as schizophrenia thought to arise in early adulthood seem to have some causes in childhood as explained later in this chapter (Luo 2021). Here, I mainly explain psychiatric disorders, as they provide us important information which contributes to a better understanding of the nature of our mind and consciousness. Then, I explain some neurodevelopmental disorders, in which attention systems are affected. Some neurodegenerative disorders are also explained.

There are various types of brain and/or mental disorders such as schizophrenia, dissociative disorders, neurodevelopmental disorders including autism spectrum disorders and attention deficit hyperactivity disorders, depression, bipolar disorders, anxiety disorders and social personality disorders such as psychopathy and sociopathy. The estimations of the lifetime prevalence of all these disorders among all populations are as follows; schizophrenia slightly less than 1%, depersonalization and derealization disorders about 1%, autism spectrum disorders 2%, attention deficit hyperactivity disorders 1.4–3%, depression 20%, bipolar disorders 2%, anxiety disorders 4–7%, post-traumatic stress disorders about 8% and psychopathy 1% (Kahn et al. 2015, Spiegel et al. 2013, Yang et al. 2023, Thapar and Cooper 2016, Wang et al. 2023, Filatova et al. 2021, Bauer 2022, Tyrer and Baldwin 2006, Baxter et al. 2013, Richter-Levin et al. 2019, De Brito et al. 2021). These mental disorders are serious problems not only for patients, their families and their colleagues, but also for the whole of human society. Thus, the development of therapy and improvement of treatment for such disorders is important. At the same time, understanding these

DOI: 10.1201/9781003518822-11

disorders contributes to the elucidation of the nature of self, mind, consciousness, emotions and the brain. Thus, I would like to explain some of them and discuss what we can learn from the studies of these disorders. In addition to the above mentioned psychiatric and neurodevelopmental disorders, there are sleep disorders, addictions, epilepsy and neurodegenerative diseases such as Alzheimer disease, Parkinson's disease, etc. Sleep disorders are very prevalent, and their prevalence could be affecting more than half of the world's population depending, on the definition (Voderholzer and Guilleminault 2012). Our brains are vulnerable to disorders or disfunctions presumably due to their complicated regulation mechanisms and functions.

Schizophrenia

Schizophrenia is a type of chronic psychiatric disorder accompanied with a combination of psychotic symptoms such as hallucinations and delusions, and motivational and cognitive dysfunctions. Symptoms of schizophrenia are divided into positive, negative and disorganization categories. Positive symptoms are behaviors and thoughts that are not normally present, such as delusions and hallucinations. The world which a schizophrenic patient sees or hears seems to be different from the world which healthy people recognize. The negative symptoms are social withdrawal, affective flattening, anhedonia, that is the inability to feel pleasure, and diminished initiative. The disorganization symptoms are various cognitive dysfunctions. Psychotic symptoms of schizophrenia usually become apparent in late adolescence or in early adulthood, although it has been considered that the origin of schizophrenia is in early life. Schizophrenia has been considered to be caused by complex interplays between genetic and environmental risk factors that influence early brain development. Schizophrenia is inheritable as shown by studies on twins, etc., and there are numerous genetic risk factors with low contributions. Individuals who experience an excess of complications during fetal life and/or around the birth have an increased risk (Kahn et al. 2015, Jauhar et al. 2022).

Hallucinations and delusions are the main positive symptoms of schizophrenia. The former is false perceptions such as patient hearing others talking about them or hearing their thoughts spoken. Delusions are persistent, bizarre and/or irrational beliefs such as a patient's belief that other people can hear his or her thoughts or that the government is monitoring all his or her actions. The contents of hallucinations and delusions in schizophrenia are in principle passive feelings, or those controlled by an external force or an agent. Subjective changes in the ownership of thinking such as thought insertion, thought withdrawal and/or thought broadcasting, and experiences that actions, sensations and/or emotions are controlled from the outside have been reported. Some delusions could be caused by hallucinations. A hallucination hearing others speak the patient's thoughts could lead to a delusion that other people know his or her thoughts. It has been considered that a primary cause of hallucination is attention to irrelevant stimuli that should be ignored such as those generated by the patient's own thoughts or actions. Misattribution of self-generated actions or thoughts to others are likely to be a main cause of hallucinations. We know our own actions and can ignore the sensory information caused by our own actions, because we can predict their outcomes. Failure in being able to make such predictions might contribute to the misattribution (Fletcher and Frith 2009, Jauhar et al. 2022). In schizophrenia patients having hallucinations and/or delusions, their own selves seem to be disturbed, disordered and/or impeded. I also would like to note that qualia of some objects in schizophrenia patients are likely to be different from those in healthy people.

Aberrant salience hypothesis of schizophrenia has been proposed, which claims that the disordered dopamine system in the ventral tegmental area causes over-attribution of meaning and motivational values to irrelevant environmental events. Some of the dopamine neurons in the ventral tegmental area encode motivational values for positive outcomes, while others respond to salient but aversive stimuli as explained in chapter eight. fMRI studies reported that schizophrenia patients show reduced activation in the ventral tegmental area and the nucleus accumbens in the basal ganglia upon presentation of reward-predicting stimuli; and have exaggerated responses to some neutral stimuli compared to non-patients. Drugs which would increase dopamine levels such as amphetamine and cocaine could cause psychosis, and antagonists of dopamine D_2 receptors have been used in antipsychotic medication. There are also evidences that dopamine synthesis and release capacity are elevated in patients with schizophrenia. Dopamine systems are explained in chapter nine. Thus, the dysfunction of dopamine systems has been proposed to be a main cause of psychosis in schizophrenia, although it might not explain all the symptoms of schizophrenia and there could be other causes too. The implication of NMDA-type glutamate receptors in schizophrenia has also been proposed mainly based on reports that non-competitive antagonists of NMDA receptors, such as, ketamine or phencyclidine induce delusions and hallucinations (Howes and Nour 2016, Jauhar et al. 2022).

Animal models of schizophrenia have been generated, which are classified into four types based on induction methods—developmental, drug-induced, through lesion and genetic manipulation. Most rodent models show behavioral phenotypes that resemble positive-like symptoms of schizophrenia, but fewer models show phenotypes analogous to negative and/or cognitive symptoms. In the developmental rodent models, manipulation of the environment such as post-weaning social isolation, drug administration during the sensitive perinatal period, etc. have been used. In the pharmacological rodent models, drugs such as amphetamine or phencyclidine have been administered to rodents. In the lesion model, neonatal lesion of the ventral hippocampus has been performed, which causes developmental alteration of neuronal integrity in the prefrontal cortices. Several types of transgenic mice have been used as genetic manipulation models of schizophrenia, such as, those with inducible and/or partial mutations of disrupted-in-schizophrenia 1 gene; various types of transgenic mice with alteration of neuregulin-1 gene; and naturally occurring dysbindin or reelin mutant mice. These mutant mice have been used in various experiments including behavioral and pharmacological studies, and are useful to better understand schizophrenia, to develop effective drugs and/or to improve the therapy (Jones et al. 2011). However, it is to be noted that human schizophrenia are heterogenous, and that each animal model could be only a partial model and does not show every phenotype of schizophrenia.

I would like to note again that patients with schizophrenia exhibit subjective changes in the ownership of self and misattribution of self-generated actions or thoughts to those coming from the outside. Thus, metacognitive function and awareness, two essential functions of consciousness are seriously affected in schizophrenia. I also note that malfunction of the dopamine system is critically involved in schizophrenia.

Dissociative Disorders

Dissociative disorders are characterized by disruption and/or discontinuity of consciousness, memory, self-identity, perception, etc., and include dissociative identity disorders, dissociative amnesia including dissociative fugue, depersonalization derealization disorders, etc. (Spiegel et al. 2013).

A person with the dissociative identity disorder has two or more distinct identities, personal states or selves, each of which has their own perceiving and thinking patterns, and also memory. These states recurrently take control of the person's behavior. A subpopulation of patients report that they feel or have felt that they are possessed by an external agent such as a spirit, a power, a deity or another person, which seems to be somewhat similar to thought insertion in schizophrenia. A childhood experience of physical and/or sexual abuse could be a cause of dissociative identity disorders (Spiegel et al. 2013).

The predominant symptom of dissociative amnesia is the inability to recall important personal information in a few episodes which are usually traumatic or stressful, but there are cases in which the memory of everyday occurrences cannot be recalled. The predominant disturbance of dissociative fugue is sudden, unexpected travel away from a home or a work place with inability to recall one's past and with confusion about one's identity. Here, a sense of personal continuity based on the maintenance of a consistent stream of memory is disrupted (Spiegel et al. 2013).

Depersonalization and derealization disorders are characterized by persistent feelings of detachment from one's self, and by those of unreality about the outside world, respectively. In the former, a patient experiences feeling of detachment from one's own mental processes, self or body, as if one is an outside observer (Spiegel et al. 2013, Yang et al. 2023).

Symptoms of dissociative disorders such as change of personality or self in the dissociative identity disorder, inability to recall memory in dissociative amnesia, and an outsider's view in the depersonalization disorder, remind me of dream experiences during REM sleeps, which are explained in chapter ten. Similar neuronal processing in both dreams and some dissociative disorders may be taking place. In any case, in dissociative disorders, the consistent unity of self and/or consciousness is disrupted, and it seems that the center of awareness and metacognitive processing is changed. The neuroscientific bases of dissociative disorders are enigmatic, presumably because of the subjective nature of their symptoms and of the lack of appropriate animal models.

Neurodevelopmental Disorders

Neurodevelopmental disorders are the early onset psychiatric and/or neurological disorders observed mainly in children, but they may have lifelong influences, and include autism spectrum disorders, attention deficit hyperactivity disorders, specific learning disorders, motor disorders, etc. (Thapar et al. 2017). The following two sections explain autism spectrum disorders and attention deficit hyperactivity disorders.

Autism Spectrum Disorders

Autism spectrum disorders are heterogenous neurodevelopmental disorders, ranging from very mild to severe, and show an early onset of social communication deficits and restricted repetitive behaviors. Social communication deficits include deficits in social–emotional reciprocity, such as failure of normal back-and-forth conversation, reduced sharing of interests or emotions, etc. Deficits in non-verbal communicative behaviors such as abnormalities in eye contact, body language, etc., and deficits in developing, maintaining and understanding relationships with others, are also observed. Restricted repetitive patterns of behaviors and interests include stereotyped or repetitive motor movements, use of restricted objects or speech, insistence on sameness such as extreme distress at small changes, difficulties with transitions and rigid thinking patterns. Very restricted and fixated interests that are abnormal in intensity or focus, hyper-reactivity or hypo-reactivity to sensory inputs, or unusual interests in sensory inputs from the environment such as indifference to pain or temperature, and adverse responses to specific sounds, have also been reported. It was estimated that around 80% of risks to contract autism spectrum disorders are heritable, although there are non-genetic risk factors. The risk to have the autism spectrum disorder is 3–4-times higher in boys than girls. Autism spectrum disorders are not considered as a focal impairment in a specific brain region or a neuronal system, but considered as a condition resulting from the overall brain reorganization (Lord et al. 2018).

Asperger syndrome and high functioning autism with normal cognitive functions are known. Asperger syndrome is accompanied by social interaction impairment, behavioral and interest restriction, normal cognitive functioning and the absence of significant general delay in language, whereas high functioning autism is accompanied with average or above average intellectual abilities. The distinction between the two have been debated. In any case, both show some autistic tendencies with normal cognitive functions (de Giambattista et al. 2018). The savant syndrome is a rare extraordinary condition in which a person showing autistic problems has some genius abilities, which are often linked to massive memory. A reported savant could name all the area codes and major city zip codes of the United States, and another showed extraordinal calculation ability such that he answered how many seconds he had lived at the age of 70 within 90 seconds: the answer was 2,210,500,800 seconds (Treffert 2009).

Animal models of autism spectrum disorders have been developed, which are classified as genetic models, copy number variation models, idiopathic models and environmental models. Genetic models include transgenic or knockout mice of synaptic adhesion proteins, other synaptic proteins, etc. which are affected in some patients. Copy number alterations, which are observed in some patients, have been reproduced in some mice models. Inbred strains of mice and rats showing autism-related social impairments and repetitive behaviors, have been regarded as models of idiopathic autism. Environmental models are produced by affecting new born rodents directly, or through the mother using infectious/inflammatory means or specific chemicals. These animal models have been used in behavioral tests, the examination of brain structures and functions, testing effects of drugs and other treatments (Wang et al. 2023). However, it is to be noted that autism spectrum disorders are heterogenous and likely to have multiple causes. Thus, a single animal model cannot cover all aspects of the disorders. In addition, most animals show phenotypes other than autism-related ones, making analyses of relationships between autism spectrum disorders and molecular and neuronal abnormalities complicated.

I wonder if restricted interests and insistence on sameness in autism spectral disorders are based on difficulties in changing attended objects, and/or biased or unusual selection of attended objects. In an autistic savant, very high mental concentration with the exclusion of most of sensory input signals, which would be based on the extraordinary high activities in restricted neuronal assemblies, might enable an incredible performance in restricted information processing. An ordinary person usually takes care of the surroundings to avoid dangers or to get possible merits, and does not completely concentrate on one thing. On the other hand, some autistic people might concentrate on an object much more strongly, and in a few cases such as a savant, the intense concentration to a restricted object or processing might produce very high quality outcomes. It is said that about 10% of autistic persons show some savant skills (Treffer 2009).

Attention Deficit Hyperactivity Disorders

Attention deficit hyperactivity disorders are heterogenous neurodevelopmental disorders characterized with inappropriate inattention, motor hyperactivity and impulsivity. The examples of inattentive symptoms are making careless mistakes, difficulty in maintaining attention on tasks, losing things, forgetting daily activities, etc., whereas those of hyperactivity and impulsivity are leaving the seat in situations when staying on a seat is expected, running about when not appropriate, difficulty in waiting one's turn, talking excessively, etc. Attention deficit hyperactivity disorders are more common in boys than girls with a ratio of 3–4 to 1. They are highly heritable and multifactorial with the contribution of multiple genes and also non-genetic factors. The onset age of attention deficit hyperactivity disorders is before twelve years of age. The symptoms tend to become unclear with advancing age, but some have difficulties as adults, those with inattentive symptoms in particular. Individuals with attention deficit hyperactivity disorders differ from each other in their core symptom combinations, levels of impairment and comorbidities (Thapar and Cooper 2016).

Animal models of attention deficit hyperactivity disorders include genetic models such as spontaneously hypertensive rats, dopamine transporter knock-out mice, SNAP-25 deficient mutant mice, nicotinic receptor knock-out mice, and tachykinin-1 receptor knock-out mice, and chemically induced models in which prenatal or early postnatal exposure to ethanol, nicotine, etc. is applied. They have been used for various researches, which have suggested involvement of dopamine, norepinephrine and serotonin systems in the disorders, although it was noted that animal models cannot truly reflect the human disorders (Russell 2011).

I think that in attention deficit hyperactivity disorders, most symptoms are primarily based on the difficulty of keeping attention to certain objects, and/or based on attention shifting away easily from attended objects, which seems to be in contrast to restricted interests and insistence on sameness in autism spectral disorders. In both disorders, attention systems seem to be not normally controlled, and the directions of alteration are opposite between the two disorders, although their high comorbidity has been reported. Drugs enhancing function of the norepinephrine system, which is involved in attention, have been prescribed to some patients with attention deficit hyperactivity disorders (Thapar and Cooper 2016).

Genetic Variations Related to Schizophrenia and Autism Spectrum Disorders

Large scale genome-wide association study identified 108 genetic loci that were associated with schizophrenia (Schizophrenia Working Group of the Psychiatric Genomics Consortium 2014). The associations were enriched among genes expressed in the brain including those coding dopamine D_2 receptor, proteins working in glutamatergic synapses, voltage-gated calcium ion channels, etc., and also among genes expressed in the immune system. It was suggested that most associated variants exert their effects through alteration of gene expression rather than that of a protein structure. Regarding autism spectrum disorders, large scale whole exome sequencing (Iossifov et al. 2014) have been performed and genetic risk factors have been studied. The contribution of each gene alteration to the risk to have disorders is relatively small (Lord et al. 2018, Jauhar et al. 2022).

Depression

Major depressive disorders are heterogeneous psychiatric disorders which are characterized by depressed mood, anhedonia, diminished interests, impaired cognitive functions and vegetative symptoms such as disturbed sleep and/or appetite. They are considered multifactorial disorders with various causes and triggers such as genetic susceptibility, stress, and other pathological processes such as inflammation. Major depressive disorders occur about twice as often in women than in men, and affect one out of 5–6 humans in their lifetimes (Otte et al. 2016, Filatova et al. 2021), and about a half of them experience multiple occurrences (Monroe and Harkness 2022).

The involvement of malfunction in the brain structures, such as the hippocampus, the amygdala, the prefrontal cortex, the hypothalamus, etc. and that in the neuromodulator systems such as the serotonin, norepinephrine systems, etc. in major depressive disorders, have been suggested by clinical studies and researches using animal models; although a number of other brain regions such as the cingulate cortex, the insular cortex and the nucleus accumbens, etc., the hypothalamic–pituitary–adrenal axis, and molecules such as stress hormone cortisol and brain-derived neurotrophic factor, etc., have also been suggested to contribute to the occurrence. Drugs targeting the monoamine systems such as selective serotonin reuptake inhibitors and serotonin–norepinephrine reuptake inhibitors, have been prescribed as antidepressants. The appearance of the drug's effect takes weeks, and neuronal and/or synaptic plasticity have been considered to be induced by the drugs; although the exact mechanisms of how antidepressants exert their effects remain enigmatic. Physical neuromodulations using repetitive transcranial magnetic stimulation, low-field magnetic stimulation, transcranial direct current stimulation, stimulation of the vagal nerve, deep brain electrical stimulation, etc. have also been applied to some patients (Otte et al. 2016, Filatova et al. 2021).

Anxiety Disorders

Patients with generalized anxiety disorders have unfocused anxiety that is not connected to recent stressful events. This disorder is twice as common in women as in men. Generalized anxiety disorder is characterized by feelings of threat, restlessness, irritability, sleep disturbance, tension, and is also accompanied by palpitation, dry mouth, trembling and/or sweating. Genetic studies suggest that generalized anxiety disorder and major depression could have a common genetic basis. The involvement of the serotonin system and the GABA system in generalized anxiety disorders have been suggested. Selective serotonin reuptake inhibitors, serotonin and norepinephrine reuptake inhibitors, and benzodiazepines binding to ionotropic GABA receptor channels, have been prescribed to some patients (Tyrer and Baldwin 2006).

In both major depressive disorders and generalized anxiety disorders, the emotion control systems and the neuromodulator systems are affected. Qualia of some objects felt by patients seem to be different from those of healthy people, because of anhedonia in depression, feeling of threat in anxiety disorders, etc. Different emotional states seem to affect general feelings, perception and awareness.

Bipolar Disorders

Patients with bipolar disorders show intensive mood swings including extreme highs and extreme lows. These extreme highs are called manic (very intense) or hypomanic (less intense) episodes including increased energy, elevated or irritable moods, little need for sleep, impulsive behaviors, etc. The extreme lows are called depressive episodes such as feelings of sadness, hopelessness and/or fatigue, and difficulty in sleeping. Patients usually show symptoms in their late teens through their early forties, and the initial episode is depression rather than mania or hypomania in most cases (Bauer 2022).

Bipolar disorder is highly heritable and mostly attributable to common variants with small effects, although environmental factors, including childhood adversity, could also contribute to its occurrence. A number of risk genes have been identified, which include those coding calcium ion channels and a molecule binding to sodium ion channels, etc. Lithium carbonate has been prescribed to patients with bipolar disorders (Harrison et al. 2018). Although the exact mechanisms of the drug function are unclear, lithium ion is known to affect calcium signaling through sodium and calcium ions exchanger, etc.

The feelings of self-esteem are contrasting in the two moods of the bipolar disorder. It is high in manic states and low in depressive states, implying the subjective self is felt or recognized differently with altered metacognitive functions. How mood swings take place in bipolar disorders is unclear; although the appropriate control of emotion systems seems to be disturbed. In the two moods, subjective qualia of a certain object seem to be different.

Post-Traumatic Stress Disorders

Post-traumatic stress disorder (PTSD) is caused by an event containing a physical threat to oneself or others that induces intense fear, helplessness or horror (Pitman et al. 2012, Richter-Levin et al. 2019, Maercker et al. 2022). It has been reported that patients with PTSD have intrusive memories of the traumatic event, avoid reminders of such memories, and experience emotional numbing and hyperarousal. They show heightened emotional reactivities to trauma-related cues, exaggerated startle, and increased sensitivity to stimuli. They also show heightened autonomic reactivity such as heart rate and skin conductance. Complex PTSD, which is caused by sustained or multiple exposures to trauma, was adopted as a new diagnosis recently (Maercker et al. 2022). Patients with complex PTSD show chronic and pervasive disturbances in emotion regulation, self-identity and relationships with others, in addition to core symptoms of PTSD.

fMRI studies reported that patients with PTSD show altered activities in the amygdala, the ventral medial prefrontal cortex, the dorsal anterior cingulate cortex, the insular cortex and the hippocampus, which are involved in the control of emotions as explained in chapter eight. An exaggerated activation of the amygdala in response to trauma-related or generic stimuli was reported. The dorsal anterior cingulate cortex shows enhanced activity during fear conditioning, etc., and the insular cortex shows increased activity to fearful or painful stimuli, etc. On the other hand, some parts of the ventral medial prefrontal cortex show decreased activation during tasks using trauma-related or trauma-unrelated stimuli. It has been proposed that decreased activities of the ventral medial prefrontal cortex fail to suppress the activities of the amygdala, which would cause attentional bias to threats resulting in increased fear responses, etc. The alterations of the neuronal systems related to amines, peptides and steroids have been suggested to be involved in PTSD (Pitman et al. 2012).

Animal models of PTSD have been generated and utilized to better understand PTSD, to develop and/or to test drugs which can be used for the therapy of PTSD (Pitman et al. 2012, Richter-Levin et al. 2019). Fear conditioning, exposure to a predator or predator odor, restraint stress or immobilization stress, submersion stress, etc. have been used as models of traumatic events, which would cause PTSD-like symptoms in animals. However, there are problems. One is that researchers need to consider ethical issues to avoid unnecessary suffering of animals. The other is that a trauma by itself is not sufficient to induce PTSD. In western populations, the life-time prevalence of severely stressful events is as high as 75–80%, but it was estimated that only about 10% of the population suffer from clinically relevant PTSD. Thus, the inclusion of exposure to risk factors, in addition to the exposure to the trauma itself, and/or inclusion of an individual profiling analysis that would enable the identification of affected individuals within the trauma exposed population, seem to be required. It has been suggested that some genetic factors and distal life experiences can be risk factors for PTSD development (Richter-Levin et al. 2019).

Psychopathy

Psychopathic individuals lack empathy, guilt and/or remorse, are callous, arrogant, deceitful, manipulative, impulsive and/or irresponsive, etc. In such persons, fMRI studies found reduced responses in the amygdala and cortical regions, such as the anterior insular and the ventromedial prefrontal cortices, which are implicated in responding to emotional stimuli during various emotional and empathy tasks by processing fear and/or pain. There are both genetic and environmental risk factors for psychopathy, and gene–environment interactions seem to exist. Psychopathy is about twice more prevalent in men than in women. Candidate risk genes include those involved in the serotonin and the oxytocin systems, which can be related to reduced emotional reactivity and attachment to others as explained in chapters eight and nine (De Brito et al. 2021).

The emotions of psychopathic individuals are likely to be different from those of other people. One's self, mind, awareness and metacognition are largely influenced by emotions. These situations remind me of the difficulty to understand or make a guess about the subjective selves and/or minds of other people with the theory of mind.

Epilepsy

Epilepsy is a type of brain diseases which is characterized by seizures, but not all seizures are due to epilepsy: there are also febrile seizures and drug induced seizures. Epilepsy is defined as two unprovoked seizures occurring more than twenty-four hours apart or a single unprovoked seizure with high recurrence risk (Manford 2017, Thijs et al. 2019). There are different types of epilepsy with different characteristics of seizure and different causes. Seizures are classified by the onset as focal, generalized, or unknown. The level of awareness subdivides focal seizures in those with retained awareness and impaired awareness. The causes of epilepsy are divided into six categories: genetic, structural, metabolic, infectious, immune and unknown. In generalized epilepsies, the epileptogenic networks are widely distributed, which include the bilateral thalamocortical networks. In focal epilepsies, the epileptogenic networks involve the cerebral cortical and/or the subcortical structures of one hemisphere in most cases (Thijs et al. 2019).

The imbalance of excitatory and inhibitory activities in neuronal circuits of the brain has been considered to be the main cause for the development of epilepsy. The imbalance in epileptogenic networks is not necessarily an increase of excitation or a decrease of inhibition; an aberrant increase in inhibition can also be pro-epileptogenic in some cases. Most generalized epilepsies are thought to have a genetic basis, whereas focal epilepsies were thought to be mostly underlined by structural cerebral abnormalities, although inherited and *de-novo* genetic mutations have been found in some focal epilepsies. Genetic mutations implicated in epilepsies are genes coding for ion channels; neuronal receptors, transcription factors and enzymes. The genetic causes are likely to be complex, involving contributions from multiple genes (Thijs et al. 2019).

During some seizures one seems to lose consciousness, and patients do not report 30–50% of their seizures. Under-reporting of seizures could be caused either by that the patients are not aware of their seizures at the time of occurrence, or that they are aware at a certain point during the seizure, but did not report them later because of poor or disturbed memory (Detyniecki and Blumenfeld 2014). Although strong neuronal activities in certain neuronal circuits in the brain seem to be related to awareness, excessive neuronal activities in wider neuronal regions in the brain during a seizure might lead to a loss of consciousness. If so, it can be suggested that not strong activities of substantial neuronal population by themselves are related to awareness, but that focused salient activities of neuronal assemblies among relatively silent surrounding neurons are likely to be important for awareness. Alternatively, if some patients with epilepsy retain subjective consciousness during certain periods of the seizure, but such subjective experiences cannot be retained, it is suggested that strong neuronal activities by themselves could give rise to some kinds of subjective experiences.

Excitation and Inhibition Balance

As explained in chapter three, there are excitatory and inhibitory neurons in the brain, and inhibitory neurons play important roles including lateral inhibition, contributing to contrast enhancement, and antagonistic inhibition. The impairment of excitation and inhibition balance seems to be a main cause of epileptic seizure as explained in a previous section, and that has also been suggested to be implicated in some other psychiatric disorders (Yizhar et al. 2011, Sawada et al. 2020).

Not only the malfunction of excitation and inhibition balance of the glutamate and GABA systems, but also that of more complicated functional balances of the control systems of attention and emotions, which are regulated by neuromodulator molecules, seem to be involved in various psychiatric and neurodevelopmental disorders. Aberrant salience hypothesis of schizophrenia claims that the disordered dopamine system causes over-attribution to the meaning and motivational values of irrelevant events resulting in hallucinations.

Very restricted and fixated interests in autism spectrum disorders could be caused by unusually persistent attention to certain objects. In attention deficit hyperactivity disorders, maintaining attention seems to be difficult. In these neurodevelopmental disorders the control of attention is not ordinary. In major depressive disorders, patients report emotional defects such as anhedonia, and also disturbed sleep, etc., which are regulated by neuromodulators. Patients with bipolar disorders show intensive mood swings including extreme highs and extreme lows, in which the appropriate control of emotions by the neuromodulator systems seem to be disturbed.

The regulation of neuronal activities by neuromodulators are not simply excitatory or inhibitory, and more complicated than those by ionotropic neurotransmitter systems, because there are multiple GPCRs for each neuromodulator molecule and they work in both directions depending on the types of coupled G protein such as Gs, Gi or Gq. Nevertheless, they work in concert and balance neuronal activities in the brain of a healthy individual. Norepinephrine is implicated in the control of attention. It has been considered that norepinephrine enhances activities of a neuronal assembly coding a salient object, while suppressing background neuronal activities in the surroundings, which should work to enhance the contrast of input signals using both excitatory and inhibitory actions as explained in chapters five and nine. Such a function of the norepinephrine system supports attention to an object or a particular information processing, by inhibiting neuronal activities related to other objects, etc. A disturbance or imbalance of such systems or functions would alter attention processes and priority judgements in the metacognition system, and would contribute to the occurrence of some symptoms of psychiatric or neurodevelopmental disorders.

Dementia

Dementia is chronic or progressive dysfunction of the cerebral cortical and the subcortical functions accompanied with cognitive decline. These cognitive changes are commonly accompanied by disturbances of one's mood, behavior and personality. Dementia should be characterized as a syndrome rather than as one particular disorder. Dementia affects about 5–7% of the elderly population over age 65 years. Different forms of dementia have been distinguished—Alzheimer's disease, dementia with Lewy bodies, frontotemporal dementia, dementia secondary to diseases, etc. (Ritchie and Lovestone 2002, Gale et al. 2018).

Alzheimer's disease is the most common neurodegenerative disease, it has a prevalence of about 5% of all individuals of age sixty-five years and above, and up to 30% in those over the age of eighty-five. Patients at early stages may show subtle forgetfulness and may show irritability, apathy and/or low mood. With the advancement of the disease, the brain MRI may show atrophy of the medial temporal lobe involving the hippocampus. Dementia with Lewy bodies seems to be the second most common neurodegenerative dementia. Its core clinical features are fluctuating cognition with pronounced variations in attention and alertness, recurrent well-formed visual hallucinations and some symptoms of Parkinson's disease, etc. (Gale et al. 2018). Lewy bodies are intraneuronal inclusion bodies composed of aggregated synuclein protein that are found in the substantia nigra of patients with Parkinson's diseases, etc. (Ritchie and Lovestone 2002). In both Alzheimer's disease and dementia with Lewy bodies, neurodegeneration affecting structures of the brain takes place. Alzheimer's disease, Parkinson's disease are explained further in the following sections together with Huntington's disease and prion disease.

Alzheimer's Disease

As written in the previous section, Alzheimer's disease is the most common neurodegenerative disease and a main cause of dementia. In a patient with

Alzheimer's disease, memory loss is followed by the impairment of other cognitive and intellectual capabilities, and a patient gradually shows difficulty in performing daily activities. Defects in synaptic and neuronal functions followed by neuronal death, which is accompanied with atrophy of cerebral cortices, take place in a patient (Luo 2021).

In the cerebral cortices of patients with Alzheimer's disease, two pathological features are observed. One is extracellular amyloid plaques, and another is numerous abnormal intraneuronal fibrils called neurofibrillary tangles. Amyloid plaques consist mostly of 39–43-amino-acids (mainly 40 and 42) peptides called amyloid β (Aβ), while neurofibrillary plaques consist of aggregates of hyper-phosphorylated microtubule-binding protein tau (Alberts et al. 2023).

Aβ is a part of the extracellular domain of a transmembrane protein called amyloid precursor protein, whose physiological function has not been clarified. Aβ is processed from amyloid precursor protein by two proteases named β-secretase and γ-secretase. Different lengths of Aβ are produced, as the cleavage site of γ-secretase is not fixed. γ-secretase is actually a protein complex which includes the transmembrane protein named presenilin. Mutations in amyloid precursor protein or presenilin have been reported in patients with early-onset familial Alzheimer's disease, although they are in a minority (1–2%) in all patients with Alzheimer's disease. Aβ deposits are toxic to neurons. However, the severity of symptoms does not correlate with the density of amyloid plaques. In addition, it has also been reported that diffuse oligomeric forms of Aβ are highly toxic to neurons, suggesting that non-aggregated Aβ may be potent Alzheimer's disease promoting agents. The most important genetic risk factor for the late onset Alzheimer's disease is the gene encoding apolipoprotein-E (ApoE), which is involved in lipid transport and metabolism. ApoE binds to Aβ and is present in amyloid plaques. The involvement of microglia in Alzheimer's disease has also been proposed (Lane et al. 2017, Luo 2021, Scheltens et al. 2021).

Parkinson's Disease

Parkinson's disease is the second most common neurodegenerative disease. The patients have difficulty in initiation of movement, and show shaking, rigidity, slowness and walking difficulty. Non-motor features of the disease include olfactory dysfunction, cognitive impairment, psychiatric symptoms, sleep disorders, autonomic dysfunctions, pain and fatigue. In Parkinson's disease, the death of dopamine neurons in the substantia nigra pars compacta sending outputs to the striatum takes place, which reduces the dopamine level in the striatum and causes hypo-activation of the direct pathway and hyper-activation of the indirect pathway in the basal ganglia. These result in the excessive activation of inhibitory outputs from the basal ganglia to the thalamus, and inhibit initiation of movements (Kalia and Lang 2015, Luo 2021). These basal ganglia pathways are explained in more detail in chapters four, eight and nine.

In the brains of Parkinson's disease patients, intracellular inclusions called Lewy bodies are found in neurons. The major component of Lewy bodies is a presynaptic protein called α-synuclein, whose mutation or overproduction leads to aggregation, which seems to cause the disease. The involvement of dysfunctions of mitochondria and lysosomes in Parkinson's disease has also been reported. The administration of L-dopa, which is a precursor of dopamine, and physical stimulation such as deep brain electrical stimulation of a part of the basal ganglia pathways, have been used for the therapy of Parkinson's disease (Kalia and Lang 2015, Luo 2021). Physical brain stimulations are explained in chapter nine.

Huntington's Disease

Most patients of Huntington's disease show depression and mood swings in the early stage followed by abnormal movements. Symptoms of the disease include chorea, dystonia, incoordination, cognitive decline, etc. Huntington's disease is caused by alterations in a single gene coding huntingtin, which is a widely expressed protein. In the gene, extension of the repeat of cytidine-adenosine-guanosine (CAG) trinucleotides sequence which codes amino-acid glutamine takes place, that results in an extended poly-glutamine repeat near the N-terminus of huntingtin. Huntingtin proteins with long ploy-glutamine repeat form aggregates, which accumulate in the inclusion bodies within a neuron. The inclusion bodies also contain some proteins interacting with altered huntingtin, resulting in the impairment of normal functions of neurons and in neuronal death, which is conspicuous in the striatum. The atrophy of the striatum and preferential effects on the indirect pathway of basal ganglia (Figure 3 in Chapter 8) in particular, have been considered to be the primary cause of characteristic excessive movements of Huntington's disease (Walker 2007, Luo 2021).

Prion Disease

Neurodegenerative diseases caused by proteinaceous infections particles (prions) have been reported. These prion diseases are accompanied with aggregation of misfolded proteins, which are induced by the infected conformation-altered proteins. The pathogenesis that is extracellular aggregation of proteins causes neurodegeneration, is similar to that of Alzheimer's disease. Three examples of prion diseases are known. The first one is caused by scrapie protein which infects sheep and goats. The second one is a human disease called kuru, which occurred in the tribes of Papua New Guinea who practiced ritual cannibalism. The third is a rare inherited disease called Creutzfeldt-Jacob disease. All of them are associated with massive neuronal death that makes sponge-like holes in the brain, and behavioral abnormalities, motor dysfunction, cognitive impairment, ataxia, etc. occur (Sigurdson et al. 2019, Luo 2021).

As described above, in neurodegenerative diseases abnormal aggregations of mis-folded proteins or cleaved fragments take place. They are toxic to neurons and cause neuronal degeneration in some regions of the brain. Affected regions are different among diseases, which seem to cause different symptoms. The reason why different brain regions are preferentially affected in each disease is unclear. It might be ascribed to different expression patterns of various proteins in neurons in different regions, which interact with the altered pathogenic proteins (Luo 2021). Most neurodegenerative diseases are accompanied with deteriorated cognitive functions such as dementia, which are consistent with the view that substantial numbers of neurons in some brain areas are indispensable for our normal cognitive activities.

Summary and Conclusion

Psychiatric disorders such as schizophrenia, dissociative disorders, depression, bipolar disorders, anxiety disorders, and neurodevelopmental disorders such as autism spectrum disorders and attention deficit hyperactivity disorders, have been reported in significant populations, and their understanding contributes to elucidation of self, mind, consciousness and emotions.

Schizophrenia is a chronic psychiatric disorder accompanied with psychotic symptoms such as hallucinations, delusions, cognitive dysfunctions, etc. with heterogeneous genetic and neurobiological backgrounds. Hallucinations are false perceptions and delusions are bizarre beliefs. It has been reported that some schizophrenia patients experience subjective changes in the ownership of thinking such as thought insertions, and have feelings that their actions, sensations and/or emotions are controlled from the outside. It has been suggested that a primary cause of hallucination in schizophrenia is attention to irrelevant stimuli that should be ignored such as those generated by the patient's own thoughts or actions. The world which a patient sees or hears seems to be different from the world which healthy people recognize. Aberrant salience hypothesis of schizophrenia claims that the disordered dopamine system in the ventral tegmental area causes over-attribution of meaning and motivational values to irrelevant environmental events. Drugs which increase dopamine levels, such as amphetamine cause psychosis, and antagonists of dopamine D_2 receptors have been used in antipsychotic medication, although the malfunction of the dopamine system cannot explain all the symptoms of schizophrenia. Metacognitive function including self-recognition and awareness, two essential functions of consciousness, are seriously affected in schizophrenia.

Dissociative disorders are characterized by disruption of consciousness, memory, self-identity, perception, etc., and include dissociative identity disorders, dissociative amnesia, depersonalization derealization disorders. A person with dissociative identity disorders has two or more distinct identities, personal states or selves, each of which has their own perceiving and thinking patterns. These states recurrently take control of the person's behavior. In dissociative disorders the consistent unity of self and/or consciousness is disrupted, implying that the self, or the center of awareness and metacognitive processing can be changed in certain conditions.

Major depressive disorders are characterized by depressed mood, anhedonia, diminished interests, impaired cognitive functions, etc., and generalized anxiety disorders are characterized by unfocused anxiety that is not connected to recent stressful events. It has been suggested that generalized anxiety disorder and major depression could have common genetic bases. In both major depressive disorders and generalized anxiety disorders, the emotion control systems and the neuromodulator systems are affected. Patients with bipolar disorder show mood swings including extreme highs and extreme lows, suggesting that the appropriate control of emotions is disturbed. Self-esteem is contrasting in the two moods of the bipolar disorder, implying that the subjective self is felt or recognized in different ways with altered metacognitive functions in the different moods.

Autism spectrum disorders are heterogenous neurodevelopmental disorders, whose symptoms include early-onset social communication deficits, restricted repetitive behaviors, insistence on sameness, rigid thinking patterns, very restricted and fixated interests, hyper-reactivity or hypo-reactivity to sensory inputs, etc. Restricted interests and insistence on sameness in autism spectral disorders might be based on difficulties to change attended objects and/or on a biased selection of attended objects. Attention deficit hyperactivity disorders are heterogenous neurodevelopmental disorders characterized with inappropriate inattention, motor hyperactivity and impulsivity, which might be based on the difficulty to keep the attention on certain objects. This situation appears in contrast to restricted interests and insistence on sameness in autism spectral disorders. In both disorders, attention seems to be not controlled normally.

Altered excitation and inhibition balance has been suggested to be implicated in psychiatric disorders. Not only disturbed balances of the glutamate and GABA systems, but also those of control systems of attention and emotions, which are regulated by neuromodulators, seem to be involved. Aberrant salience hypothesis of schizophrenia claims that disordered dopamine system causes over-attribution of meaning and motivational values to irrelevant events, and it has been suggested that misattribution of self-generated actions or thoughts to others are likely to be a main cause of hallucination. Very restricted and fixated interests in autism spectrum disorders could be caused by unusually persistent attention to certain objects, and in attention deficit hyperactivity disorders difficulty to maintain attention takes place. Patients with bipolar disorders show intensive mood swings including extreme highs and extreme lows, in which the appropriate control of emotions by neuromodulator systems seems to be disturbed. Thus, the disturbed functions of neuromodulator systems involved in the control of attention and emotions, seem to be involved in the occurrence of various psychiatric disorders by affecting awareness and metacognitive functions. Most of the associated genetic variants related to schizophrenia or neurodevelopmental disorders exert their effects through alteration of gene expression rather than that of a protein structure, implying that the main causes of these disorders may be the disturbed quantitative balances of the expression of various proteins, which are likely to cause altered activity balances of neuronal systems related to attention and/or emotions.

References

Alberts, B., K. Hopkin, A. Johnson, D. Morgan, K. Roberts, P. Walter et al. 2023. Essential Cell Biology. 6th Ed. Norton, New York, NY.

Bauer, M.S. 2022. Bipolar disorder. Ann. Intern. Med. 175: ITC97–ITC112.

Baxter, A.J., K.M. Scott, T. Vos and H.A. Whiteford. 2013. Global prevalence of anxiety disorders: a systematic review and meta-regression. Psychol. Med. 43: 897–910.

De Brito, A.A., A.E. Forth, A.R. Baskin-Sommers, I.A. Brazil, E.R. Kimonis, D. Pardini et al. 2021. Psychopathy. Nat. Rev. Dis. Prim. 7: 49.

de Giambattista, C., P. Ventura, P. Trerotoli, M. Margari, R. Palumbi and L. Margari. 2018. Subtyping the autism spectrum disorder: comparison of children with high functioning autism and Asperger syndrome. J. Autism Dev. Disorders 49: 138–150.

Detyniecki, K. and H. Blumenfeld. 2014. Consciousness of seizures and consciousness during seizures: Are they related? Epilepsy Behav. 30: 6–9.

Filatova, E.V., M.I. Shadrina and P.A. Slominsky. 2021. Major depression: one brain, one disease, one set of intertwined processes. Cells 10: 1283.

Fletcher, P.C. and C.D. Frith. 2009. Perceiving is believing: a Bayesian approach to explaining the positive symptoms of schizophrenia. Nat. Rev. Neurosci. 10: 48–58.

Gale, S.A., D. Acar and K.R. Daffner. 2018. Dementia. American J. Med. 131: 1161–1169

Harrison, P.J., J.R. Geddes and E.M. Tunbridge. 2018. The emerging neurobiology of bipolar disorder. Trends Neurosci. 41: 18–30.

Howes, O.D. and M.M. Nour. 2016. Dopamine and the aberrant salience hypothesis of schizophrenia. World Psychiatry. 15: 3–4.

Iossifov, I., B.J. O'Roak, S.J. Sanders, M. Ronemus, N. Krumm and D. Levy. 2014. The contribution of *de novo* coding mutations to autism spectrum disorder. Nat. 515: 216–221.

Jauhar, S., M. Johnstone and P.J. McKenna. 2022. Schizophrenia. Lancet 399: 473–486.

Jones, C.A., D.J.G. Watson and K.C.F. Fone. 2011. Animal models of schizophrenia. British J. Pharmacol. 164: 1162–1194.

Kahn, R.S., I.E. Sommer, R.M. Murray, A. Meyer-Lindenberg, D.R. Weinberger, T.D. Cannon et al. 2015. Schizophrenia. Nat. Rev. Dis. Prim. 1: 1–23.

Kalia, L.V. and A.E. Lang. 2015. Parkinson's disease. Lancet 386: 896–912.

Lane, C.A., J. Hardy and J.M. Schott. 2017. Alzheimer's disease. Eur. J. Neurol. 25: 59–70.

Lord, C., M. Elsabbagh, G. Baird and J. Veenstra-Vanderweele. 2018. Autism spectrum disorder. Lancet 392: 508–520.

Luo, L. 2021. Principles of Neurobiology. 2nd Ed. CRC Press, Boca Raton, FL.

Maercker, A., M. Cloitre, R. Bachem, Y.R. Schlumpf, B. Khoury, C. Hitchcock et al. 2022. Complex post-traumatic stress disorder. Lancet 400: 60–72.

Manford, M. 2017. Recent advances in epilepsy. J. Neurol. 264: 1811–1824.

Monroe, S.M. and K.L. Harkness. 2022. Major depression and its recurrences: life course matters. Annu. Rev. Clin. Psychol. 18: 329–357.

Otte, C., S.M. Gold, B.W. Penninx, C.M. Pariante, A. Etkin, M. Fava et al. 2016. Major depressive disorder. Nat. Rev. Dis. Primers 2: 16065.

Pitman, K.R., A.M. Rasmusson, K.C. Koenen, L.M. Shin, S.P. Orr, M.W. Gilbertson et al. 2012. Biological studies of post-traumatic stress disorder. Nat. Rev. Neurosci. 13: 769–787.

Richter-Levin, G., O. Stork and M.V. Schmidt. 2019. Animal models of PTSD: a challenge to be met. Mol. Psychiat. 24: 1135–1156.

Ritchie, K. and S. Lovestone. 2002. The dementias. Lancet 360: 1759–1766.

Russell, V.A. 2011. Overview of animal models of attention deficit hyperactivity disorder (ADHD). Curr. Protoc. Neurosci. 9: Unit 9.35.

Sawada, T., T.E. Chater, Y. Sasagawa, M. Yoshimura, N. Fujimori-Tonou, K. Tanaka et al. 2020. Developmental excitation-inhibition imbalance underlying psychoses revealed by single-cell analyses of discordant twins-derived cerebral organoids. Mol. Psychiat. 25: 2695–2711.

Scheltens, P., B. De Strooper, M. Kivipelto, H. Holstege, G. Chételat, C.E Teunissen et al. 2021. Alzheimer's disease. Lancet 397: 1577–1590.

Schizophrenia Working Group of the Psychiatric Genomics Consortium. 2014. Biological insights from 108 schizophrenia-associated genetic loci. Nat. 511: 421–427.

Sigurdson, C.J., J.C. Bartz and M. Glatzel. 2019. Cellular and molecular mechanisms of prion disease. Annu. Rev. Pathol. Mech. Dis. 14: 497–516.

Spiegel, D., R. Lewis-Fernández, R. Lanius, E. Vermetten, D. Simeon and M. Friedman. 2013. Dissociative disorders in DSM-5. Annu. Rev. Clin. Psychol. 9: 299–326.

Thapar, A. and M. Cooper. 2016. Attention deficit hyperactivity disorder. Lancet 387: 1240–1250.

Thapar, A., M.M. Cooper and R. FRCPsych. 2017. Neurodevelopmental disorders. Lancet Psychiat. 4: 339–346.

Thijs, R.D., R. Surges, T.J. O'Brien and J.W. Sander. 2019. Epilepsy in adults. Lancet 393: 689–701.
Tyrer, P. and D. Baldwin. 2006. Generalised anxiety disorder. Lancet 368: 2156–2166.
Treffert, D.A. 2009. The savant syndrome: an extraordinary condition. a synopsis: past, present, future. Phil. Trans. R. Soc. B 364: 1351–1357.
Voderholzer, U. and C. Guilleminault. 2012. Sleep disorders. Handb. Clin. Neurol. 106: 527–540.
Walker, F.O. 2007. Huntington's disease. Lancet 369: 218–328.
Wang, L., B. Wang, C. Wu, J. Wang and M. Sun. 2023. Autism spectrum disorder: neurodevelopmental risk factors, biological mechanism, and precision therapy. Int. J. Mol. Sci. 24: 1819.
Yang, J., L.S.M. Millman, A.S. David and E.C.M. Hunter. 2023. The prevalence of depersonalization-derealization disorder: a systematic review. J. Trauma Dissociation 24: 8–41.
Yizhar, O., L.E. Fenno, M. Prigge, F. Schneider, T.J. Davidson, D.J. O'Shea et al. 2011. Neocortical excitation/inhibition balance in information processing and social dysfunction. Nat. 477: 171–178.

12

Short-Term and Long-Term Merits and Demerits for Individuals, Groups and Species

Conflicts of Short-Term and Long-Term Merits and Demerits

In the majority of behavioral choices or decision making, some conflicts between merits and demerits, and/or between short-term and long-term effects arise. For example, an animal may need to choose whether it should get a small amount of food which can be accessed easily, or it should try to get a large amount of food which might be available only after a long journey. This choice, that is to be made, corresponds with the conflict between short-term and long-term merits. In another situation, an animal may need to choose whether to try to get food which can be found in a dangerous area where predators are many, which corresponds to a conflict between a merit and a demerit.

We, humans, also experience similar kinds of conflicts in every-day life, in company, in society or at national level problems. You may wonder whether you should have lunch or continue to work in order to finish a task as soon as possible. In a company there may be conflicts about how many people and how much budget should be assigned to on-going works, which are expected to provide certain immediate profits; or to the development of a new plan whose outcome is uncertain presently, but which might provide a big chance to expand profits in the future. Similarly, in academia there seem to be conflicts about the division of the scientific budget to basic and applied sciences, or science and technology, although they are highly dependent upon each other. In a nation and/or world-wide, there are conflicts about energy resources. For example, to obtain electric power, how much natural energy, nuclear power or fossil fuel should be used. The use of natural energy has been highly recommended in order to reduce the output of carbon dioxide, which elevates the temperature of the earth, and also to reduce the risk of accidents in nuclear power plants which are likely to affect the environment seriously. However, the use of natural energy may cost a lot, may not be easily expanded and it may also

DOI: 10.1201/9781003518822-12

cause some ecological problems in certain cases. Furthermore, the abrupt shut-down of thermal power plants may cause economic and/or employment problems. There are also conflicts about a nation's budget, such as, how much national bond should be issued. National bonds might be necessary for immediate use such as to provide a budget for disaster recovery or for the activation of recessed national economy, etc., but it is a debt which has to be paid in the future by younger generations.

There are actually different opinions about such social issues. It may be reasonable for people who consider such social problems seriously, to give more importance to long-term effects. Over use of national bonds and/or thermal power electricity generation are likely to increase social debts in the future. However, it is also to be noted that recession and unemployment may have a devastating effect on the society, a nation and citizens in a shorter time scale. Serious demerits such as the bankruptcy of a company and the collapse of a nation, might make possible long-term merits meaningless. Thus, both short-term and long-term effects are important, and an appropriate balance and/or adequate adjustments of each effect are required. Having enough resources to spare is likely to support activities for long-term merits.

Conflicts of Merits and Demerits Between Individuals, Groups and/or Societies

Conflicts of merits and demerits exist among individuals or groups of animals. Among rivals for getting food or a partner of the opposite gender, there will be winners and losers. There are cases in which animals give a higher priority to the family or the group to which they belong, than to their own merits. For example, a parent bird may give first priority to feeding its babies; and working bees may give first priority to supporting their colony. In some cases, there seems to be a conflict in an animal's behavioral decision making. For example, when two baby birds are in danger, as when a snake is approaching, a parent bird needs to make the decision whether it should try to save the babies, taking the risk of losing its own life. In case it does decide to help, which baby bird should be saved first, may become a conflict. As described in chapter two, rats avoid actions that harm other rats, although the harm aversion is suppressed when it can get a large amount of food by taking the action to harm another rat. To which merits should one give the first priority, depends on the situation; but it could be a difficult problem on some occasions.

In human societies, there are more difficult or complicated cases of conflict among individuals, one's family, the group or the company to which one belongs, the society, the nation and/or the whole human race. A leader of a group, which can be a company, a certain political group, a city or a nation, may be expected to give the first priority to the group, although he or she is likely to have a family and may be a member of other groups also. In such situations, a leader seems to face a conflict in making a decision regarding whose merits should be given higher priority and whether short-term effects or long-term effects should be given higher priority. In many cases, certain adjustment and/or compromises are necessary, and the leader should consider how to persuade related parties and get the agreement of each group.

In a democratic society, compromise is expected to be attempted, although it is often not an easy job. The merits for a certain group can be demerits for other groups such as rival groups; or demerits of a larger group to which the group belongs to. A political action of a nation to get its own merit may sometimes disturb merits of other nations. Action selections considering merits and demerits of other nations seem to be required, even though the decision may decrease immediate merits of the nation itself, and the adequate consideration of merits of other nations might benefit the nation through getting the trust of other nations in the future.

Decision Making in Conflicting Conditions and Development of Metacognition

As described in previous sections, decision making and/or action selection, considering short-term and long-term merits and demerits, and also those of different related parties is not easy on many occasions. There is no correct answer because of which everyone related to the issue will be satisfied in many cases. Conflicts in decision making occur in everyday life. You may have to attend to some sudden urgent work when you have an appointment with your friend or your family. Your child may cry, or an unexpected visitor might arrive, when you are cooking or talking on the phone at home. There can be more serious conflicts that require you to choose whether you save your own child or five unknown people.

I think that such conflicts in behavioral selections to obtain better outcomes, have developed our metacognitive abilities. Not only humans, but also many animals show metacognitive abilities as explained above. Thus, the development of metacognitive abilities seems to take place during animal evolution. I wonder if such development of metacognitive abilities, dealing with a number of different and possibly conflicting information, has somehow contributed to the generation of subjective consciousness, mind and self. One reason of such a speculation is that we are not aware of our automatic responses such as reflexes, etc., and also do not remember well most of actions that are conducted without any hesitation or conflicts.

Game Theory

The game theory was initially developed to deal with economic problems, and according to Cambridge Learner's Dictionary its definition is "an economic theory that tries to calculate what choices people will make, when those choices depend on the choices of others". The application of the game theory has been extended to various fields, and its more general definition is "a mathematical theory about how decisions are made in situations where one person's decision affects another, that is used in many fields such as economics, psychology, and biology". In biology the game theory has been used to study the evolution of various social behaviors such as contest behavior, cooperation, etc. (Archetti and Scheuring 2012, Leimar and McNamara 2023).

A famous topic in the game theory is prisoner's dilemma, which has been explained below (Fogel 1996). Two people A and B committed a crime, and were arrested, held in separate cells and were unable to communicate with each other. A prosecutor proposed a deal to A and B separately, and told them that their partner had also been offered the same deal. Then the prosecutor told them separately, "We have a lot of circumstantial evidence against both of you. So, if you both claim innocence, we will convict you anyway and you'll both get two years in jail. But if you will help us out by admitting your guilt and making it easier for us to convict your accomplice, we'll let you out free. And don't worry about revenge – your accomplice will be in jail for five years!" The prosecutor also added that if both A and B admit the guilt, both of A and B would be in jail for four years.

What is the better choice for A? First, let's postulate that B admits being guilty. In this case, if A also admits being guilty, A will be in jail for four years, and if A claims innocence, A will be in jail for five years. Thus, A should choose to admit guilt. Now, let's consider the case that B claims innocence. In this case, if A also claims innocence, A will be in jail for two years, but if A admits the guilt, he will be free. So, to admit guilty is a better choice. Thus, rational consideration leads to the conclusion that A should admit being guilty, in any case. However, B is also likely to think in the same way. If both A and B admit the guilt, both will be in jail for four years. Obviously, this outcome is worse than if both claim innocence, which will result in two years of jail for both. Thus, the dilemma becomes apparent. If A and B can trust each other, they could get a better outcome. Trust and cooperation between individuals can lead to good results in certain cases. This case study also makes it clear that rationally correct consideration cannot always lead to the best outcome.

Researchers have used the game theory to explain some animal behaviors including cooperation, etc., and also to find adequate strategies to obtain larger outcomes, which have provided interesting ideas, although they are inconclusive (Fogel 1996, Leimar and McNamara 2023). In human societies with multiple participants with unknown characteristics, situations can be far more complex and ambiguous, and action selection or behavioral choice is difficult in many cases.

Roles of Self-Centered Emotions in Decision Making

As explained in chapter eight, emotions are self-centered subjective experiences, and significantly influence decision making, attention and memory. Emotions are mainly triggered by external stimuli, largely influenced by one's environment, and depend on one's own experiences. Emotions like fear and reward strongly influence learning, memory and attention through activities of neuromodulator molecules in the amygdala, the nucleus accumbens, etc. as explained in chapters eight and nine. Fear, anxiety and reward consequently affect action selections, together with other emotions, such as affection, anger, disgust, etc. through neuronal activities in the prefrontal cortices, etc.

Emotions that influence decision making and action selection of animals, are considered to have evolved to merit the survival and breeding of individual animals, or the prosperity of the group to which the individual animal belongs to, as explained in chapters two and eight. Fear which is caused by factual and acute threat and anxiety which is caused by a potential and/or anticipated threat, evoke defensive behavioral responses, to avoid or to reduce harm and support the survival of the animal (Tovote et al. 2015). Anger and aggression work to secure resources such as food, territory and/or a mating partner, and to defend oneself and the family (Lischinsky and Lin 2020). The functions of disgust have been considered to expand from the original role of defending the animal's body from toxins and diseases, to roles that work in defending it from various potential threats such as individuals violating social and/or moral norms (Chapman and Anderson 2012). Pleasure or internal rewards work to promote behaviors, which contribute to the animal's survival, such as getting food, or reproduction of offsprings through mating (Berridge and Kringelbach 2015). Love, attachment and empathy also seem to contribute to the prosperity of the family or the group to which the individual animal belongs (Feldman 2017).

As mentioned above, most emotions are evoked by external stimuli coming from the nearby environment and work to support the merits or to reduce the demerits of an individual and/or closely related animals or humans. Thus, emotions are effective in a relatively small group in most cases, and may not be well suited to make decisions about issues related to merits and demerits of larger groups including the species as a whole.

Size of The World and Human Power

The size of the earth is very large for animals other than human beings. Most animals only know the limited areas where they live in, except for migratory birds, mammals, fishes, etc. which travel over relatively large areas of the earth. The influence of activities of each animal species on the environment is confined to a relatively small area, except for the influence exerted by human beings. In such situations, animals have evolved in various ways depending on the environments of the local areas they live in. Some have enhanced adaptability for the survival of individual animals improving their metacognitive activities and emotions, which fit competition and survival strategies in a relatively small group and in a restricted area. I think that human beings have also evolved similarly.

In olden days, when people knew only nearby places, their activities were limited in the area and in the effect that their actions had on the natural environment. Our emotion systems seem to have been designed to adapt to such situations. However, today human beings have very large and strong powers, which influence the environment of the earth extensively. For example, if substantial nuclear weapons would be used, very serious human and environmental problems lasting many years would occur. It is also to be noted that our daily and ordinary activities continue to produce a large amount of carbon dioxide resulting in global warming, which influences climates, the production of crops and vegetables, the sea level, etc. in the whole world. Human beings are winners in the competition with other animal species, and our population has increased. However, a further increase in the world-wide human population might cause problems in the delivery of food, water, energy, etc. With improvements in the means of transport and communication, everyone knows what is happening in each corner of the earth. In addition, economical interaction between nations such as trading and other international interactions have largely increased. In such conditions, some events in a particular nation may have a great influence on other nations. We know what is happening in other nations, which makes me feel that the world has become smaller.

Past Experience and Education

Memory of past experiences together with emotions largely influence one's decision making. However, one's past experience is relatively limited compared with various events occurring in the whole world. In general, better decision making or action selection can be made with more accurate information, which is used for metacognitive processing. Thus, providing accurate information by education, etc. is required for each individual to have improved capability to make adequate decisions.

The Emergence of Distinctive Rules in Higher Order Systems

The decision-making processes of a human group, a society, a nation, and an individual are somewhat similar in a sense that all of them have internal conflicts. Human societies have their own rules, norms, etc. which are diverse among societies and are complicated. They cannot be easily related to characteristics of individual members. By the way, the emergence of new rules at a higher level is not confined to human societies, in fact it seems to be a general principle as explained below.

There is a hierarchy in the physical world consisting of elementary particles, atoms, molecules, polymers, molecular assemblies, cells containing various molecular signaling cascades, neuronal circuits, brains, individuals, groups of individuals, etc. Each level seems to have its own laws, rules and/or phenomena which are based on the laws of a lower level. Some of the higher-level phenomena and laws can be explained by the laws of the lower level; however higher-level phenomena and laws are often brought about by complicated interactions of components of the lower level, and are often difficult to be explained by using the laws of the lower level. It seems as though new laws and rules emerge at a higher level.

A simple chemical reaction can be explained by quantum physics, but complicated chemical reactions of large molecules, such as proteins, are difficult to be explained in a similar way. Alternatively, they can be more easily understood based on the laws of various chemical reactions. Molecular signaling cascades within a cell, such as those explained in chapter three, consisting of many molecules and many chemical reactions including feedback, converging and diverging pathways, yield various outcomes depending on situations of the cascades and inputs to the system. These behaviors of signaling cascades are difficult to be explained based on specific chemical reactions, and the total grasp of the system seems to be necessary to understand its working mechanisms (e.g., Bhalla and Iyengar 1999, Kitagawa et al. 2009). Similarly, to understand working mechanisms of neural circuits and the brain, not only the cellular level understanding of neurons and synapses, but also the total understanding of their interactions is required. For example, as explained in chapter eight, neuronal systems processing and controlling various emotions are very complicated, and their total understanding has not been reached.

One theoretical reason for the difficulty of explaining higher-level phenomena and/or laws based on laws of the lower level, is that even if laws in the lower level are described in a deterministic dynamical system, outcomes in the higher level can be very variable depending on the initial values. In other words, a small change in an initial value can result in a large difference in the output, which is called chaos. The existence of chaos-like or near-chaotic phenomena have been suggested to occur in the brain (O'Byrne and Jerbi 2022).

In a human group or society, many laws, rules, norms, doctrines, etc. have emerged. In most cases they have been developed so that the group or society is stabilized and maintained by reducing confusions and conflicts within the group, and also by reducing risks to the group. These rules have been constructed by members of a group and based on characteristics of individual humans. Indeed, there are certain similarities in the rules of human groups and societies, such as, prohibition of stealing and injuring others, promotion of cooperation and obeying rules, and protection of the group or society from external enemies, etc. However, details of the rules can be significantly different and are difficult to be predicted by characteristics of each member.

The differences in functions, characteristics, environments, sizes, histories, etc. of groups, seem to contribute to diversities of detailed rules. I would like to note that a group, society, etc. may become an organization consisting of individuals, who might be regarded as components, each of them playing a restricted partial role. It seems that an organization itself has its own purposes, and the weights or importances of the organization and the members are different depending on characteristics of organizations. Recent complicated interactions among individuals, groups, societies and/or nations including trading, economics and/or military relations, together with technical improvements such as advancement of AI and communication technologies, etc., have accelerated the complication of rules of human societies.

Purposes and Means

A means easily becomes a purpose in a human society. A means is a way or an approach to accomplish a certain purpose. Human purposes such as getting wealth, honor, good reputation and/or respect in a society, might be originally the means for obtaining a high position in a group or in the society, which would merit an individual and/or the family so that their lives become more fulfilling, enjoyable and happy. However, getting wealth or honor itself might become a purpose in certain cases. A person who wished to become a leader or a politician might hope to contribute to the society at the beginning. However, for some of them to be a leader or a politician might become a purpose in itself. Laws and rules in a society should be constructed to enable fair judgement regarding conflicts, and for things to proceed smoothly. However, it could become troublesome when situations have changed. Laws and rules might come to stand by themselves and they might force members to obey them. Laws and rules are means, but they can become purposes. Readjustment of rules are required when circumstances have changed.

From an evolutional view point, emotions are considered to have been evolved to merit individual animals or a group to which one belongs, although too strong or uncontrolled emotions such as fear, anxiety or anger can disturb one's action selection and may give demerits to an individual and/or a group as described in chapter eleven. There have been thoughts in human societies that emotions should be suppressed to make adequate decisions, and in recent days some people might consider that AI or computer systems without emotions could provide fairer and more unbiased judgement. I think that these ideas have been born, partly because emotions are self-centered and may not be suited to support decision making for a large group and/or for long-term merits as discussed above. However, it is to be noted that what is the best or a better choice is often difficult to determine, mostly because of the difficulty about what should be given higher priority. Which of the short-term or long-term effects are more important, and/or which populations or groups should be given higher priority, are hard to be determined in many cases.

Judgement, Prediction and Decision Based on Facts, Reasoning, Opinion, Belief, Hope and/or Emotions

Decision making and/or action selection is made through metacognitive functions based on predicted outcomes, which are brought about by the evaluation of facts and reasoning. One's own opinions, beliefs, hopes and/or emotions also influence decision making. This scheme is not confined to individuals, but is also applied to groups, local societies, nations and international societies.

Facts, reasoning, opinions, etc. are interrelated. Reasoning is based on facts, predictions are made through various reasonings, and opinions are formed by evaluating and referring to predictions. But their relations are neither straightforward nor simple. Opinions, beliefs, hopes and emotions are likely to influence the selection of facts which should be focused upon and used for inference. Reasoning is also likely to be biased by opinions, beliefs, hopes and emotions. In addition, accuracy of facts and reasoning can be questionable in some cases as explained in following sections.

Among human individuals, societies and nations, there are various opinions and beliefs, and there are also conflicts of interests. Thus, forming a consensus among them is often difficult. However, in the present era, when human beings have great power to influence the environment of the whole world or earth, adequate and wise decision making of individuals and various societies and nations is necessary for maintaining human societies and for the survival of human beings.

Facts

You may think that to discern facts is an easy job, but I wonder if it is really that easy. You will believe that what you have seen is a fact. But as I explained in chapters five and seven, you might have illusions. When you get some information by hearing it or reading about it, the reliability of the source needs to be considered. In case a source of information is an individual, he or she may have misunderstood something; may consider a speculation as a fact; may report only a portion of an event giving a biased impression; or may simply lie. If you get information from a website or a social network service, the reliability of the source needs to be evaluated, as there are many biased and/or false contents available. I note that the development of AI has made it easy to construct realistic fabricated images and movies. Multiple independent reliable sources should be checked or cross-referred, if the information is critical. You also need to keep it in your mind that web providers may select information so that it fits your interest by referring to your past search history, resulting in biased information presentation. To obtain unbiased information about facts is not easy.

Various Ways of Reasoning

There are various ways or types of reasoning such as inference, speculation, deductive reasoning and empirical or heuristic reasoning, and each has different accuracy of estimation. According to Cambridge Learner's Dictionary inference is defined as "a belief or opinion that you develop from the information that you know", whereas speculation is "the act of guessing possible answers to a question without having enough information to be certain". Thus, inference is based on firmer or more sound information than speculation, and seems to yield more accurate estimation or prediction.

Deductive reasoning is definitive reasoning without doubt such as syllogism, which states that if A = B and B = C, then A = C. However, there are many cases when deductive reasoning cannot be applied, and in such cases empirical or heuristic reasoning is used. Established physical laws such as Newton mechanics, Maxwell equations in electromagnetism, theory of relativity, etc. that are considered true, are in fact empirical or heuristic rules. They were constructed based on observed phenomena, and verified by the confirmation of predicted phenomena. Natural sciences have developed in such a way. The observation of various phenomena and gathering facts are followed by attempts to find a natural law that underlies the phenomena, and a hypothesis about the law is constructed. Based on the hypothesis and by using logical or deductive reasoning, some predictions about experimental or observational results are made, and the accuracy of prediction is tested or examined in order to judge the correctness of the hypothesis or the law.

Empirical reasoning includes Bayesian speculation as explained in chapter six. Other speculation, based on uncertain information, is often used in our daily activities, although the accuracy of estimation or prediction can be low in some cases. I also note that even if in a situation deterministic deductive reasoning can be applied, the prediction of outcomes can be very variable depending on the initial values in some cases; in other words, chaos may occur. In addition, the selection of facts and phenomena, and empirical reasoning can be biased by one's opinions, beliefs, emotions, etc. Thus, we should consider the accuracy of reasoning, and it is also critical to distinguish facts from inference and speculation. Inferences and speculations are often claimed to be facts.

Opinion, Belief, Hope and Emotions

We have some opinions, beliefs and hopes, which have been formed through our respective experiences, and in general it seems that we tend to have strong opinions and beliefs on interested matters and/or affairs. For some people their opinions and beliefs may be more important than facts and rational reasoning. One's own opinions and beliefs significantly influence one's decision making together with one's hopes and emotions.

It is also to be noted that they affect the selection processes of information. Most people try to get information related to their own interests, and may not pay enough attention to uninterested affairs. Further, some may choose only information, which they like or want to hear, and may also conduct biased reasoning emphasizing positive information and neglecting negative ones. On the other hand, some anxious people may do the opposite in selection of information and in reasoning. In any cases, I think it is important to recognize that most of us, if not all, have some biases in the selection of information and possibly in the reasoning processes.

How We Should Judge and Decide

How we should judge and make decisions as an individual and/or a group does not seem to have a unified correct answer which can be applied to every case. Nevertheless, there are several points we should take care of, as have been mentioned above. Not only short-term but also long-term merits and demerits of individuals and various groups including companies, societies, etc. need to be considered. We should take care about the selection of facts and the accuracy of the reasoning. We need to keep it in our mind that the selection of facts and reasoning can be biased by opinions, beliefs, hopes and emotions of related people, and that what we think of as facts are not always true, and also could be inferences or speculations. Thus, the distinction among facts, reasonings and opinions is critical. Finally, I would like to note again that predictions cannot always be correct, and that there are almost always some conflicts of merits and demerits, between short-term and long-term effects, and also among individuals and groups.

I think that our emotions and attention systems tend to attach more importance to short-term effects, which might be accompanied by long-term demerits. Rational opinions that consider long-term effects, such as keeping earth warming to the minimum and abandoning nuclear weapons have been suggested. These are reasonable opinions and should be considered seriously. However, shorter-time effects cannot be ignored, because if shorter-time demerits reach a certain level, they might cause a total change in situations, such as bankruptcy of companies accompanied with increased unemployment and social anxiety, or unstable international relationships. Therefore, it seems necessary to perform adequate adjustment or balancing of short-term and long-term effects, and to conduct careful procedures to explain situations to related people and to form consensus among them.

The game theory explains that trusting others can result in getting merits in some cases, but it also mentions that the strategy of trusting others always, may have the likelihood of resulting in loss of merits or suffering because of demerits (Archetti and Scheuring 2012, Leimar and McNamara 2023). Thus, we need to take care to decide whether others can be trusted or not. Indeed, there are psychopaths and sociopaths in this world. However, we have empathy, sympathy and compassion, and we do cooperate and decide to help or support others in many cases. Altruistic behaviors are more likely to be conducted by people who have the capacity or allowance to support others, although there are many exceptional cases. As mentioned before, a rat makes a behavioral choice of not hurting another rat, but this behavior is converted when a large merit, a large amount of food in this particular case, can be obtained by doing so. However, if the rat had eaten enough before the need for the behavioral choice arose, the conversion of behavioral choice to one that harms another rat, might not occur. Thus, I wonder if wealthy or satisfied people with larger capacity and allowance could be more easily altruistic or generous to others, and that they may also be able to attach more importance to long-term effects. Thus, a fair distribution of wealth and resources among individuals, societies and nations seems to be important, although what is fair is a difficult problem and there may be diverse opinions.

Adequate adjustment of human population is an important problem, which influences demand, supply and distribution of food, water, energy and other resources in the world. Human beings have the highest brain functions and are placed on the top of the ecological pyramid. All other species on the earth are unable to threaten our survival. Only very large-scale natural disasters, including collision with an asteroid, emergence of very dangerous infectious microorganisms, and misconduct or wrong actions of some influential human groups, etc., are threats to our survival. Thus, the control of the population of human beings seems to primarily depend on ourselves. The question regarding what can be considered as being an adequate population, is related to amounts of resources such as food, water, energy, territory, etc. that are available for our use. Technical developments are expected to increase them, but the speed matters. As human beings have got a great deal of power and a huge population, the capacity of the earth to keep the environment sustainable for us, has become a serious problem.

In Japan as well as in some other countries and races, population has been decreasing. In the long-term, I myself, think that a moderate decrease in human population may not be so bad for us, as then individuals may be able to have better access to various resources and have more opportunities to spend satisfactory and enjoyable lives in principle. But in shorter time scales, a biased decrease in the population in certain areas is likely to cause serious social problems, such as difficulty in maintaining infrastructures, including delivery systems of energy, water, etc., and also of maintaining local societies. At a national level, a smaller population will decrease the economic power of the country, resulting in the reduced influence in the international society; while some other nations with increased human populations and enhanced economic powers, come to show stronger influence in the international society. Such short-term demerits are too significant to be ignored. I wonder if some changes in thoughts on individual happiness, social rules and norms, economic status, other circumstances of human societies, etc. might have caused the decrease of human population in some societies. In any case, adequate adjustment and balancing of short-term and long-term effects are difficult in general.

Finally, I would like to discuss ways of getting reliable information. Previously, mass media, such as newspapers and television, were the main providers of information about societies, nations and the world. Developments of social network services (SNS) and smartphones have changed the situations. Now many people seem to obtain more information through SNS and smartphones, and the relative importance of mass media as providers of information seems to have decreased. This situation might be problematic, as there are so many biased and false information contents available, and the distinction of facts from speculations or beliefs can be often unclear in SNS. In addition, the decrease in the social influence of reliable mass media may weaken their efforts and abilities to gather factual information because of the decrease in financial support.

I also wonder if the relatively small display sizes of smart phones might possibly interfere with deep and thoughtful understanding and consideration of various situations and problems. Not so many information contents can be presented simultaneously on a smartphone, which makes it difficult to refer to or compare related information parallelly, and might confine one's thinking to serial processing of limited information. As explained in chapter five the capacity of our working memory is limited, we cannot handle or process so much information at a time. We should also be aware that SNS and the web often provide biased information by referring to one's previous search history. The above situations might restrict one's interests and viewpoints. We need to make certain that efforts are made to obtain wide unbiased reliable information before making appropriate decisions.

Summary and Conclusion

On most occasions when we make decisions or select actions, there are some conflicts between short-term and long-term merits and demerits, and also conflicts among merits and demerits of related individuals and groups. In many cases, there is no correct answer with which everyone can be satisfied. I think that such conflicts in behavioral selections to obtain better outcomes, have developed or strengthened our metacognitive abilities, which I feel might have somehow contributed to the generation of subjective consciousness. This is because we are not aware of our automatic responses, such as reflexes, etc., and also do not remember most of actions or behaviors that are conducted without any hesitation or conflict.

Decision making and/or action selection are performed through metacognitive functions based on predicted outcomes, which are brought about by evaluation of facts and reasonings. One's own opinions, beliefs, hopes and/or emotions influence decision making. This scheme is not confined to individuals, but is also applied to groups, local societies, nations and international societies. Facts, reasoning, opinions, etc. are interrelated. Reasoning is based on facts, predictions are made through various reasonings, and opinions are formed by evaluating and referring to predictions. But their relations are neither straightforward nor simple. Opinions, beliefs, hopes and emotions are likely to influence the selection of facts which should be focused. Reasoning is also likely to be biased by opinions, beliefs, hopes and emotions. In addition, the accuracy of facts and reasoning can be questionable in some cases. Among humans, societies and nations, various opinions, beliefs, etc. abound, as do conflicts of interests. Thus, consensus formation among related individuals and/or groups is usually difficult. However, in the present era, when human beings have large power to influence the environment of the whole world, adequate and wise decision making of individuals, various societies and nations, is necessary to maintain human societies.

Emotions that influence decision making and action selection of animals are considered to have evolved to merit the survival and breeding of individual animals or the prosperity of the group to which the individual belongs. Most emotions are evoked by external stimuli from the environment and work to support merits or to reduce the demerits of an individual and/or closely related animals or humans. Thus, it seems that emotions are effective in a relatively small group in most cases, and they might not be suited well to consider merits and demerits of large groups, societies, nations and/or the whole world. I also think that our emotions and attention systems tend to attach more importance to short-term effects, which could cause long-term demerits in some cases. Rational and reasonable opinions considering long-term effects, such as, that earth warming should be kept to a minimum, have been claimed. However, shorter-time effects cannot be ignored, because if shorter-time demerits reach a certain level, they might cause total changes in situations, such as social anxiety and unstable international relationships. Therefore, adequate adjustment or balancing of short-term and long-term effects and careful procedures of explanation of situations, consensus formation, action plan formation and its implementation, seem to be required.

The game theory explains that trusting others can result in getting merits in some cases, but it also maintains that the strategy of always trusting others, may result in loss of merits. Thus, we need to be sure whether others can be trusted or not. However, we also have empathy, and we do cooperate and make decisions to help or support others in many cases. Altruistic behaviors are more likely to be conducted by people who have enough capacity or allowance to support others.

References

Archetti, M. and I. Scheuring. 2012. Review: Game theory of public goods in one-shot social dilemmas without assortment. J. Theor. Biol. 299: 9–20.

Berridge, K.C. and M.L. Kringelbach. 2015. Pleasure systems in the brain. Neuron 86: 646–664.

Bhalla, U.S. and R. Iyengar. 1999. Emergent properties of networks of biological signaling pathways. Sci. 283: 381–387.

Cambridge Learner's Dictionary. 4th Ed. 2012 Cambridge Univ. Press, Cambridge, U.K.

Chapman, H.A. and A.K. Anderson. 2012. Understanding disgust. Ann. N. Y. Acad. Sci. 1251: 62–76.

Feldman, R. 2017. The neurobiology of human attachments. Trends Cogn. Sci. 21: 80–99.

Fogel, D.B. 1996. Special issue on the prisoner's dilemma. BioSystems 37: 1–2.

Kitagawa, Y., T. Hirano and S. Kawaguchi. 2009. Prediction and validation of a mechanism to control the threshold for inhibitory synaptic plasticity. Mol. Systems. Biol. 5: 280.

Leimar, O. and J.M. McNamara. 2023. Game theory in biology: 50 years and onwards. Philos. Trans. R. Soc. Lond. B Biol. Sci. 378: 20210509.

Lischinsky, J.E. and D. Lin. 2020. Neural mechanisms of aggression across species. Nat. Neurosci. 23: 1317–1328.

O'Byrne, J. and K. Jerbi. 2022. How critical is brain criticality? Trends Neurosci. 45: 820–837.

Tovote. P., J.P. Fadok and A. Lüthi. 2015. Neuronal circuits for fear and anxiety. Nat. Rev. Neurosci. 16: 317–331.

13

Reconsideration of Self, Mind and Consciousness

What is The Brain for, and What is Consciousness?

As described mainly in chapter two, I think that the brains of animals have evolved as sophisticated information processing systems that serve for the survival of an individual animal and for the prosperity of its offsprings; both of which eventually contribute to the survival of the species. During evolution, recognition of objects; attention to critical and/or important objects which should be focused on; metacognitive functions to understand each situation, to predict possible outcomes of a behavior; to evaluate and to compare merits and demerits, and short-term and long-term effects, seem to have developed in order to make better or more adequate decisions and action selection. Emotions also seem to have developed to support metacognitive functions through providing a respective value to each object or affair, and also by changing the working conditions of global neuronal systems in the brain, so that they adapt to the environment or the situation using neuromodulator molecules as explained in chapters eight and nine.

Recognition of objects and attention to certain objects seem to be required to process information related to the objects consistently and without confusion. Recognized and attended objects seem to be accompanied with salient sensory signals or are chosen by top-down attention, and the neuronal assemblies coding those objects seem to show intense activities. They might also suppress activities of neuronal assemblies coding incompatible or unrelated objects (Figure 1). These effects seem to be manifested by neuronal responses to noisy visual images (Figure 2 of Chapter 1) and those with multiple overlapped objects (Figure 1 of Chapter 5), and by our recognition of ambiguous figures (Figure 4 of Chapter 5). I wonder if objective neuronal activities related to recognition and attention are accompanied with subjective awareness in most cases. Subjective awareness seems to emerge with the improvement of the ability to recognize objects and to select information using attention systems, and strong neuronal activities of certain neuronal populations seem to underly the awareness that is a main function of consciousness.

DOI: 10.1201/9781003518822-13

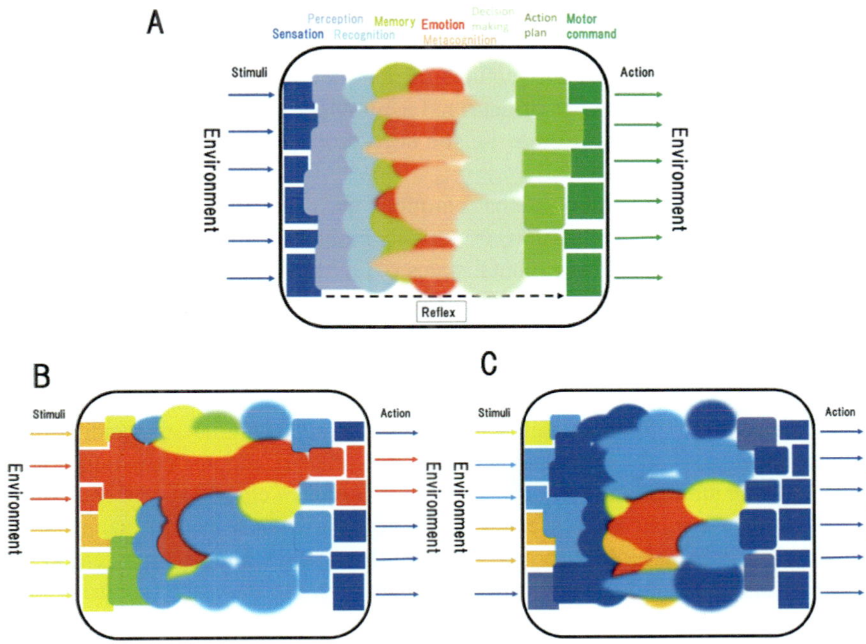

Figure 1. Conceptualistic presentation of neuronal assemblies involved in sensation, perception, recognition, memory, emotion, metacognitive function, decision making, formation of action plan and motor command. Squares and circles indicate neuronal assemblies. Squares represent neuronal assemblies coding information close to either inputs or outputs, whereas circles represent those coding higher order information. There are overlaps and interactions including mutual inhibition among neuronal assemblies. Neuronal assemblies are not fixed, and member neurons can be changed and their borders are not clear in those involved in processing higher cognitive information and emotions in particular. In A, colors of circles and squares reflect color-tagged functions. In B and C, warm colors indicate high neuronal activities and cold colors indicate low neuronal activities. B shows activities of neuronal assemblies in an awaked state, and C shows those during a REM-sleep dream. Neuromodulator molecules change global states of the brain and activities of neuronal assemblies. In an awaked states (B) salient sensory signals are perceived, recognized, processed together with stored information in a metacognitive system, and motor commands are generated. Such an information flow is schematically presented with red-colored circles and squares. In a REM-sleep dream (C) sensory inputs are neither perceived nor recognized, even if they stimulate receptor cells and induce neuronal activities in lower sensory systems. Intrinsic neuronal activities occurring in higher brain regions (red and orange circles) generate the dream. Active neuronal assemblies in an awaked state (B) and in a REM-sleep dream (C) are different. Thus, centers or most active regions of the brain, which might underly awareness and subjective self or mind, seem to change from time to time in different states or situations.

Metacognitive functions or information processings of the brain, including recognition of oneself, contribute to better decision making and appropriate action selection through understanding situations, predicting possibilities in the future, evaluating and comparing possible outcomes. In the above comparison processes, there are some conflicts between merits and demerits, and short-term and long-term effects. Emotions influence metacognitive processes and decision making by adding certain self-centered values to particular objects or affairs. I wonder if conflicts and involvement of emotions in objective metacognitive processes may be related to the emergence of subjective consciousness, self and mind, that is the main body of thinking and feeling. During processing conflicts and strong emotions, I think that there are various intense interactions of neuronal activities of both excitation and inhibition among a number of neuronal assemblies whose components may be partly shared or overlapped (Figure 1), and such processes seem to be consciously recognized.

In chapter eight, I mentioned that emotions are accompanied with behavioral changes, including facial expressions in not only humans but also animals, by which one can make a guess regarding mental states of an individual. Understanding other people's emotions and/or thoughts helps action selection or behavioral decisions. For example, if one notices that another individual is angry, one can leave the latter to prevent its attack. If one finds that another individual shows facial expression of pleasure or disgust after eating a particular food, one can know that the food is tasteful or toxic, respectively. Such an ability to guess or to understand another individual's mind, in other words the theory of mind as explained in chapter one, might have been enabled or strengthened by one's own experiences of emotions, and could have made it possible to relate various events or situations to their respective values. Such mechanisms might have contributed to the birth of subjective feelings, emotions, consciousness and mind. As we cannot see our own faces without a mirror, in a situation when a particular emotion occurs, subjective emotions might be necessary to understand the status of our own mind and/or the brain.

During dreams in REM sleep, we have bizarre subjective experiences in which we think and behave differently from the way we do in awakened states. In dreams, we ourselves, seem to be different from the way we are in the awakened states. As described in chapter ten, REM sleep is accompanied with neuronal activities partly similar to, but distinct from those in awakened states (Figure 1, also Figure 1 of Chapter 10). Certain neuronal activities underly dreams, and amines and other neuromodulator molecules control non-REM and REM sleeps and awakened states. Thus, global neuronal activity patterns in the brain can be extensively altered by neuromodulator molecules, and subjective ourselves and consciousness may also be fundamentally changed in accordance with the global brain states. Global changes of working states of diverse neuronal circuits by neuromodulators can add significant flexibility and complexity to functions of the brain.

Consciousness of AI

As discussed in chapter seven, our brains and computer systems have some similarities in their overall architecture (Figure 1 of Chapter 7). Both possess central information processing units which control input and output devices, and different types of memory systems. Both the brain and AI implemented in a computer system have learning abilities, and the recent improvement of AI seems to make it difficult to distinguish the communication responses of AI from those of a human, when the responder cannot be seen. Thus, some people may feel that AI might have mind and consciousness. However, it is to be noted that AI's responses have been learnt from many previously successful or satisfactory communications with humans, and that they are likely to be exemplary and somewhat formal. However, responses of

AI could be biased by materials used for learning and by the program. Many other people may think that AI does not have a mind and consciousness, and that AI may be something like a philosophical zombie. I have repeatedly written that subjective mind and consciousness are strictly personal, and that any person cannot directly access another person's mind and consciousness, including those of AI. Therefore, I think that it is impossible to declare whether AI has subjective consciousness or not.

However, I note that AI is not a life and does not have its own intrinsic purpose, which is a critical difference from the brains of animals. Emotion systems in the brain seem to have evolved to merit the survival of individuals and the prosperity of the species as described in chapter eight. As far as I know, no information processing systems similar to emotional systems of animals have been implemented in AI, and I also doubt the appropriateness of the implementation, as emotions are essentially self-centered. Emotions strongly affect our awareness, attention and decision making, and its absence in AI makes me feel that AI is not likely to have a subjective mind and consciousness. Even if AI has them, I think that they are likely to be quite different from ours. I also note that control systems of emotions are very complicated with extensive interactions among many neuronal areas including loop circuits using slowly working neuromodulator molecules as explained in chapters eight and nine, and they have not yet been well clarified.

In chapter twelve, I have noted that conflicts in decision making or action selection might have been related to the emergence of subjective consciousness, including awareness and metacognitive activities, because we are not aware of and do not remember most of our automatic responses. Some conflicts in decision making in an individual or in a group or society can be very serious and no correct answer may exist. In such cases, we seem to evaluate each information, to get a reasonable prediction, to compare merits and demerits, to think of short-term and long-term effects, and to consider the effects of each related individual and/or group. These decision-making processes are likely to be more conscious and remembered better as they are more difficult and laborious. Thus, I wonder if problem solving in conflicting situations might have somehow contributed to the birth of subjective consciousness.

If I adopt the above speculation, providing conflicting problem tasks to AI might result in the emergence of subjective consciousness in AI. If we force AI to give an answer to a question that does not have any correct answers, what will happen? AI might continue the never-ending process of finding an answer or might provide an answer based on learned information and a calculated evaluation of predicted outcomes, which would largely depend on the program and the information that has been previously provided to AI. The latter case in which AI makes a decision, can be dangerous for us, because AI's decision would be based on previous information which may not be appropriate for the provided question in the new situation, or may be based on inappropriate judgement criteria. Serious and conflicting questions are difficult to answer, mostly because of the difficulty in value judgements among incomparable matters and/or affairs such as the lives of different people, and of uncertainty about short-term and long-term effects, etc. To answer such questions judgement criteria are needed, but general judgement criteria which can be applied to all or most cases, and with which most people can agree or admit, are hard to be constructed, if not impossible. I think that it is better or more convincing for us to restrict the use of AI, so that it provides us with high quality processed information and calculated predictions together with the source information, and we humans decide what to do by utilizing or referring to the information provided by AI.

Self, Mind and Consciousness in Different States

We experience different states of self, mind and consciousness in sleep and dreams as explained in chapter ten. In some non-REM sleep, we seem to lose consciousness, although we can have short and simple dreams in non-REM sleeps. During REM sleep we often have long, vivid, complicated and story-like dreams. Such dreams are often bizarre, and distorted in time and space, and in most of such dreams we do not recognize that we are dreaming and accept the narrative as it is. These REM sleep dreams are often accompanied with some emotions, and we also perform some action selections in them.

Recognition of situations and decision making or action selection in story-like dreams is usually very different from those we would make in awakened states, and I sometimes wonder why and how I accepted the strange situations without doubt and behaved in strange ways in my dreams. I feel that the self in such a dream is very different from myself in awakened states, nevertheless the same my brain should underly the selves and consciousness in both states. The global neuronal activity patterns of the brain influenced by neuromodulator molecules, seem to be largely different between the awakened states and dreams. Centers and ways of information processing seem to be also different (Figure 1). One obvious difference between awakened states and dreams is that the information comes only from memory in dreams, whereas we have sensory inputs in an awakened state. Lack of salient sensory inputs might be the cause of unstable and unrealistic stories of dreams, together with weakened attention systems in the brain. In dreams, protagonists, places and situations change abruptly, which might be caused by the travelling of neuronal activities among neuronal assemblies, coding relatively unrelated objects. Such situations might be brought about by decreased attention keeping neuronal activities.

In some psychiatric disorders such as schizophrenia and dissociative disorders, the self of a patient seems to be affected as explained in chapter eleven. In schizophrenia subjective changes in the ownership of thinking such as thought insertion, thought withdrawal and thought broadcasting, and experiences that actions, sensations and/or emotions are controlled from the outside of selves, have been reported. Thus, such a patient with schizophrenia might not have confidence on his or her own self and/or will. Schizophrenia patients also have hallucinations and delusions, which are false perceptions, and bizarre and/or irrational beliefs, respectively. The world which a patient sees seems to be different from the world which healthy people recognize.

A person with dissociative identity disorder has two or more distinct identities, personal states or selves, each of which has its own perceiving and thinking patterns. These states recurrently take control of the person's behavior. Thus, the patients have multiple selves. A patient with another dissociative disorder dissociative fugue, shows sudden unexpected travel away from a home or a work place with the inability to recall one's past. Here, a sense of personal continuity based on the maintenance of a consistent stream of memory seems to be disrupted. A person with depersonalization disorder, experiences feelings of detachment from one's own mental processes, self or body, as if one is an outside observer. Thus, in dissociative disorders, the consistent unity of self is disrupted. Self may not be so solid or a firm existence.

Awareness and attention systems can be affected in some neurodevelopmental disorders as explained in chapter eleven. Some patients with autism spectrum disorders show insistence on sameness, difficulties with transitions, very restricted and fixated interests. Restricted interests and insistence on sameness might be based on difficulties to change attended objects. On the other hand, attention deficit hyperactivity disorders are characterized with inappropriate inattention, motor hyperactivity and impulsivity. Most symptoms of attention deficit hyperactivity disorders might be primarily based on difficulties in keeping attention on certain objects, in contrast to restricted interests and insistence on sameness in autism spectral disorders. In both disorders one's attention system seems to be affected, although the directions are opposite. Excitation and inhibition balance in neuronal circuits in the brain is likely to be disturbed in these disorders. As mentioned above, some patients with schizophrenia experience hallucination, which is a disturbed perception of the environment. Thus, awareness an aspect of consciousness seems to be altered in some psychiatric and neurodevelopmental disorders.

Objective Centers of Metacognition

As described in the previous section and in chapters ten and eleven, the center of perception, recognition and/or metacognitive functions, such as thinking and action selection, is not likely to be confined to a single existence, but rather there seem to be multiple centers in a single person or in a single brain (Figure 1). If so, what kinds of neuronal activities underly such multiple centers?

I speculate that strong activities of some neuronal assemblies are likely to underly perception and recognition of sensory inputs or remembered objects, and that travelling waves of neuronal activities among neuronal assemblies coding related objects or affairs, are the basis of metacognitive functions, including thinking and decision making (Figure 1). I suppose that centers of such neuronal activities are not single but multiple, are not fixed and can move within certain brain areas. I also think that centers are not likely to be confined to a particular region but are distributed in relatively broader neuronal networks, including multiple areas such as the cerebral cortices, the basal ganglia, the thalamus, the amygdala, the hippocampus, etc. Such centers of neuronal activities are determined by the interaction of salient sensory inputs, memory related signals, emotion-related signals and/or top-down attention signals. Neuromodulator molecules can change the global states of the brain and also centers of neuronal activities. Centers or most active regions of the brain, which might underly awareness and subjective self or mind, may also change in different states or in different situations. I also think that any neuronal assemblies are not fixed, and member neurons can be changed to a certain extent. There seem to be overlaps and also interactions including lateral inhibition among neuronal assemblies. The above activities of neuronal assemblies seem to be objective phenomena underlying subjective awareness and metacognitive functions.

Unity of Consciousness and Self

As has been discussed above, there seem to be multiple different states of self, mind and consciousness. In story-like dreams in REM sleeps self, awareness and metacognitive activities work differently from the way they do in awakened states (Figures 1), implying that multiple selves can exist in an individual. Patients with dissociative identity disorder show multiple selves, which recurrently take control of consciousness and behaviors, and those with dissociative fugue show disrupted personal continuity based on maintenance of a consistent stream of memory. Patients with schizophrenia are reported to experience subjective changes in the ownership of thinking and also feel that their actions, sensations and/or emotions are controlled from outside. Thus, there could be multiple selves, and identity and ownership of selves might not be so firm or stable in certain situations.

There is another example casting doubt about the unity of consciousness. That is a report about a patient with the split brain (Gazzaniga 1998, de Haan et al. 2020). A patient with epilepsy underwent surgery to cut the corpus callosum (Figure 2B of Chapter 4), which is the main pathway connecting the right and left hemisphere of the cerebral cortices. Thus, information exchange between the two hemispheres was largely interrupted. To this patient, several different pictures were presented to the right and left visual fields. For example, in the right visual field, a large picture of a house and trees covered with snow, and four small pictures showing a bicycle, a scoop, a broom and an axe were presented. In the left visual field, a large picture of a chicken leg, and four small pictures showing a toaster, an apple, a hammer and a chicken were presented. Then, the patient was asked to point a small figure related to a large figure using a right finger for pictures in the right visual field, and then a left finger for figures in the left visual field. The right finger pointed to a figure of the scoop, and the left one pointed out the figure of a chicken. Next, the patient was asked why he had selected each small figure. He answered that he chose the scoop because it was necessary to remove snow, but he could not explain why he had chosen the picture of the chicken with his left finger even though he had made the correct choice.

How were these responses brought about? I need to explain three basic neuroscientific facts before discussing the interpretation of the above experimental results. First, the visual sensory signals of the right visual field go to the left hemisphere through the optic chiasma, and those of the left visual field go to the right hemisphere (Figure 6 of Chapter 4). Second, the right and left fingers are controlled by the left and right motor cortices, respectively. Third, information processing centers for languages are located in the left hemisphere of the cerebral cortex in most people (Hickok and Poeppel 2007, Friederici 2011). Thus, it seems that the sensory signals coming from the right visual field went to the left hemisphere, processed, and the correct answer was pointed using a right finger, and also the reason of selection was verbally explained. On the other hand, the sensory information coming from the left visual field seems to have been processed in the right hemisphere and the correct answer was pointed with a left finger, but the reason could not be explained verbally, because the right hemisphere lacked the necessary language processing center for verbal responses.

In this case how should we consider the consciousness of the right hemisphere? Although verbal expression could not be provided by the right hemisphere, it performed the recognition of objects and also performed complicated metacognitive information processing. Thus, I would like to think that the right hemisphere by itself has its own consciousness and mind. If so, it seems that there can be two selves in one brain. In ordinary people, information exchange between the two hemispheres coordinates their activities, and supports united decision making and action selection. The split-brain case reminds me of the alternate neuronal activities of the cerebral hemispheres of cetaceans during unihemispheric sleep, which was explained in chapter ten. Although, I do not know the consciousness and the mind of cetaceans, they might have two relatively independent minds in certain situations.

The above experiment on a person with the split brain also reminds me that there may be cases in which conscious experiences cannot be reported. The causes of the inability may not be confined to the difficulty of making verbal reports. There might be cases that conscious experiences cannot be remembered. A white dream whose contents cannot be recalled may be an example. Thus, to have a conscious experience, to remember it and to report verbally may not be the same.

Summary and Conclusion

The brains of animals seem to have evolved as sophisticated information processing systems, that work for the survival of an individual and the prosperity of its offsprings; both of which eventually contribute to the survival of the species. During the course, recognition of objects, attention to critical objects, and metacognitive functions to understand each situation, to predict possible outcomes of a behavior, to evaluate and compare short-term and long-term merits and demerits, have developed in order to make better decisions. Emotions have also developed to support metacognitive functions. Recognition of objects and attention to certain objects seem to be required to process information related to attended objects consistently without confusion. Recognized and attended objects are accompanied with salient sensory signals or are chosen by top-down signals, and neuronal assemblies coding those objects seem to show intense activities and might also suppress activities of neuronal assemblies coding incompatible or unrelated objects. Subjective awareness seems to emerge during the improvement of the abilities to recognize objects and to select information using attention systems. Strong neuronal activities of certain neuronal populations seem to underly awareness, that is a main function of consciousness. I also wonder if subjective experiences of emotions might have emerged to better understand emotions and thoughts of other individuals.

There seem to be different states of self, mind and consciousness. In story-like dreams in REM sleeps, self, awareness and metacognitive activities work differently from those in awakened states, implying that multiple selves can exist in an individual. Patients with dissociative identity disorder show multiple selves which recurrently take control of consciousness and behaviors. Thus, the unity of self, mind and consciousness is questioned, and there could be multiple selves. Another example suggesting multiple selves in an individual, is a case study on a patient with the split-brain. Each cerebral hemisphere of the split brain seems to have abilities to recognize various objects and to perform complicated metacognitive information processing leading to a correct answer independent of another hemisphere. Thus, it seems that multiple consciousness and mind can exist in one human brain.

References

de Haan, E.H.F., P.M. Corballis, S.A. Hillyard, C.A. Marzi, A. Seth, V.A.F. Lamme et al. 2020. Split-brain: what we know now and why this is important for understanding consciousness. Neuropsychol. Rev. 30: 224–233.

Friederici, A.D. 2011. The brain basis of language processing: from structure to function. Physiol. Rev. 91: 1357–1392.

Hickok, G. and D. Poeppel. 2007. The cortical organization of speech processing. Nat. Rev. Neurosci. 8: 393–402.

Gazzaniga, M.S. 1998. The split brain revisited. Sci. Am. 279: 50–55.

14

Concluding Remarks

Aims of This Book

In the preface I have mentioned that I aim to provide a unified view of self, mind and consciousness based on neuroscientific and biological findings. Thus, I have drawn some general conclusions by relating various information and by giving a right position to each information in a grand framework. While writing this book I have also kept the following two points in mind. One is to include related topics as far as I can so that a bird's eye view of the whole field can be provided, and another is to keep descriptions accurate so that facts, inference, speculation and my opinion can be distinguished.

The reason why I have attempted to accomplish the above is as follows. Although there are many splendid and detailed review papers and textbooks, they deal with relatively specific topics, or each theme tends to be explained parallelly. In particular, as far as I know, there have been no or very few intensive trials to relate all information about molecular and cellular neuroscience, neuronal circuits functions, higher brain functions, emotions, sleeps and dreams, psychiatric disorders, evolutions, social aspects of animals and AI, to subjective self, mind and consciousness. I think it is important to have an image of the whole field, to clarify the relations among various topics and respective information, and to locate each topic and information in the right position in a large framework. When I was performing experimental research on specific topics, that were synaptic functions, synaptic plasticity and functions of the cerebellum, I was unable to try the above attempt, because there were too many things to do for rigorous verification of specific conclusions in each research.

I have found that to accomplish the above aims is difficult, and I am not sure how much I have succeeded. The difficulties are that lives and brains are complicated systems with many exceptions, and that there are many unknown and/or unresolved points which may be clarified in the future. I also have found that it is impossible to explain each topic in detail because of the space and also of my limited ability. Thus, I am not confident whether my explanations are sufficient. I do think that there are many important facts, inferences, notions that I have not written about in this book. I would like to ask interested readers to read referenced reviews, etc. for further information. To write a simple, easily understandable meaningful general

DOI: 10.1201/9781003518822-14

comments by referring to or not neglecting exceptional cases, is also a very difficult task. Indeed, what was thought as an exceptional case has become a rather general phenomena in a number of scientific themes. Thus, future neuroscientific research will certainly add or revise some of points that I have written here.

Main Conclusions

Now, I would like to state the main message of this book.

1. Neuronal activities in the brain, which are based on biophysical and biochemical reactions, underly the subjective self, mind, consciousness and emotions.
2. The main functional components of the brain are neurons and synapses, which transmit information by electrical and chemical means.
3. The main original role of the nervous system seems to be to mediate information from sensory organs to muscles, so that an animal can respond adequately to the stimulus coming from the environment.
4. The brains of animal seem to have evolved as sophisticated information processing systems, contributing to the survival of an individual and the prosperity of its offsprings, which eventually contribute to the survival of the species.
5. Sensation, perception and recognition seem to be processed with different but overlapping neuronal components in the brain.
6. Memories and emotions seem to have evolved to merit the survival of an individual and the prosperity of its offsprings.
7. Synaptic plasticity, that is the alteration of the efficacy of synaptic transmission depending on the neuronal activities, underlies learning and memory.
8. Subjective mind, consciousness and emotions are strictly personal, and cannot be directly accessed by others.
9. Consciousness has multiple meanings, which are arousal, awareness and metacognitive functions.
10. Strong activities of neuronal assemblies seem to underly awareness, and travelling waves of such neuronal activities might be bases of metacognitive functions. Both bottom-up and top-down attentions influence awareness.
11. Neuronal assemblies coding specific objects or memories are possibly not fixed with alteration of member neurons, and there seem to be also overlaps among neuronal assemblies.
12. Working memory seems to be a center of metacognitive processing.
13. Loops, divergence, convergence, lateral inhibitions and also volume transmission with diffusive neuromodulators are prevalent among neuronal circuits in the brain.
14. The global brain states such as awakened states and sleep states are controlled and altered by neuromodulator molecules such as amines, acetylcholine and orexin.

15. The brain and a computer system or AI, have similarities as information processing systems implemented with learning ability and multiple types of memory, although there are differences. One difference being that only the brain has inborn purposes as a part of life.
16. To speed up information processing, perception, recognition and decision making with comparably slowly working components, neuronal circuits in the brain seem to have developed characteristic ways or methods such as contrast enhancement using lateral inhibition and filling-in of lacking portions, at the sacrifice of accuracy.
17. Emotions seem to be processed and controlled by large-scale neuronal networks including various loops and mutual interactions in broad brain regions using not only one to one synaptic transmission but also neuromodulator molecules.
18. Emotions influence learning and memory, and the effects seem to be mainly mediated by amine neuromodulators such as dopamine and norepinephrine on synaptic plasticity.
19. Ourselves and consciousness seem to be different in awakened states and in story-like dreams during REM sleeps.
20. In some psychiatric disorders, such as dissociative identity disorder or schizophrenia, multiple selves can exist or the ownership of self or mind seem to be disturbed, respectively.
21. Excitation and inhibition balance is critical for normal functions of the brain, and its disturbances may cause some brain disorders.
22. Subjective selves and objective brain centers of recognition and metacognitive functions seem not to be confined to unity, and can be changed in different situations.
23. I wonder if conflicts of short-term and long-term merits and demerits, and conflicts among individuals and groups in decision making, might be related to the birth of subjective metacognitive functions.
24. I also wonder if subjective emotions and feelings might have developed to improve the understanding of other people's mind or thoughts, which would support the well-being and survival of an individual and/or a group during evolution.

The above points include facts, inferences, speculations and opinions, but I think that they are and will be valid in principle, partly because they are rather general and abstract statements which seem to be applicable in most cases. Future neuroscientific research will add more information, clarify some concepts more precisely and/or add some relatively minor revisions, except for points 22, 23 and 24, which are my speculations and opinions.

Critical Remarks on Main Conclusions

Specific critical remarks for each of above main statements are as follows.

1. This comment about biophysical and biochemical bases of subjective mind, etc. is based on facts, my own experiences and verbal reports of humans. However, subjective phenomena and existences are personal, and one's mind, consciousness and subjective emotions cannot be directly accessed by others as written in point 8. Thus, strictly speaking this statement is not a fact but should be regarded as an inference or an opinion.
2. The statement about neurons and synapses is a fact, but cells other than neurons such as glial cells are also likely to play important roles in the information processing in the brain, which will be clarified further by future studies.

3,4. Comments about the roles and evolution of the nervous systems and brains are inferences and/or opinions, although I think that most of researchers are likely to agree.

5. The comment about processing sensory information is an inference, although the definition or distinction between perception and recognition is not very clear.
6. The comment about the roles of memory and emotions is an abstract inference, a speculation or an opinion, and there may be more specific and detailed roles of each memory and each emotion.
7. The comment about synaptic plasticity is an inference, although additional mechanisms such as plasticity of intrinsic activities of neurons and altered functions of glial cell might contribute to learning and memory.
8. This statement about subjective experience and existence is an inference or an opinion based on my own experience and other peoples' written idea.
9. The comment about consciousness is a fact based on previous publications, but there is ambiguity about the definition of metacognitive functions.
10. The comments about activities of neuronal assemblies and attentions include inferences and speculations. As stated above, metacognitive functions have not been clearly defined. In addition, travelling waves of neuronal activities are an abstract concept without clear definition and details are unknown.
11. The comment about neuronal assemblies is an inference and a speculation. Neuronal assemblies coding specific objects or memories are an abstract concept and not well defined.
12. The comment about working memory is an opinion. There are ambiguities in the definitions of working memory and metacognitive functions.
13. The comment about neuronal circuits and neuromodulators is a fact, although details of some neuronal loops, divergence, convergence and lateral inhibitions have not been defined or clarified.
14. The comment about global brain states is a fact, although global brain states are abstract concepts.

15. The comment regarding the comparison of the brain and a computer system is an opinion. The evaluation of degree of similarities and differences may depend on individuals.
16. The comment about the information processing characteristics of the brain is an inference, a speculation and/or an opinion. How the filling-in is achieved by neuronal circuits is enigmatic.
17. The comments about control systems of emotions include facts and inferences.
18. The comments about the effects of emotions and neuromodulators on learning and memory, include facts and inferences. There might be several ways through which emotions and neuromodulators influence learning and memory.
19. The comment about selves in awakened states and in REM-sleep dreams is an inference and/or a speculation based on my own experience and other peoples' verbal reports.
20. The comment about selves in psychiatric disorders is an inference and/or a speculation based on scientific literature referring to other peoples' verbal reports.
21. The comment on excitation and inhibition balance is an inference. I note that there are multiple types of excitation and inhibition balances.
22. The comment about multiple selves is an inference and an opinion based on my own experience, other peoples' verbal reports and scientific literatures.

23,24. The comments about origins and purposes of the subjective mind, consciousness and emotions are my speculations and/or opinions without firm factual bases. They have come to my mind while writing this book. I welcome free future discussion and exchange of opinions regarding these points.

Unresolved Questions and Future Problems

Now I would like to list some of unresolved critical neuroscientific questions which I hope will be elucidated in the future.

1. The integration of synaptic inputs and the effects of neuromodulators on a single neuron need to be clarified further. In particular, how a particular neuron codes specific objects, affairs and/or memories is a critical question. This is because a neuron, itself, is a complicated system receiving various and many (1,000–100,000) synaptic inputs and is influenced by multiple neuromodulators.
2. The interactions of multiple neuromodulators and synaptic plasticity mechanisms in a single neuron together with underlying functions and interactions of intracellular molecular signaling cascades, need to be clarified further.
3. In some neurons, multiple neurotransmitters and/or neuromodulators are secreted. How each secretion is regulated and functional roles of these neurons, are worth studying further to understand complexity of a single neuron. In addition, which neuron secretes which neuropeptide has not been elucidated well. The clarification of this question together with that of roles of various neuropeptides are important.

4. The roles of various intrinsically active neurons in respective neuronal networks should be clarified further.
5. How the brain and neuronal circuits are formed during development needs to be clarified further. In particular, what processes are genetically determined and what processes depend on neuronal activities should be clarified. Such neurodevelopmental studies would contribute towards understanding neurodevelopmental disorders better.
6. Related to the above, how the characteristics of each type of various neurons in the brain are determined and how specific synaptic connections are formed are important questions which have not been precisely understood, and should be investigated further. The clarification of these points would also contribute to better understanding of neurodevelopmental disorders, etc.
7. Neuronal assemblies coding particular objects, affairs and/or memory need to be clarified more precisely. At present, the borders of neuronal assemblies regarding the activity level of each neuron are unclear or not consistently defined. I also think that it is useful to know more about the characteristics of each member neuron of an assembly, such as, whether its outputs are excitatory or inhibitory; what is the identity of its input and output neurons; the receptors of neurotransmitter and neuromodulators expressed on the neuron; the stability as a member of an assembly and other assemblies to which it belongs, etc. Such information should be interrelated and should be considered to better understand functions of a neuronal assembly. I also would like to know further about the distribution of neurons in a functional neuronal assembly, in particular, I am curious whether the locations of member neurons in an assembly are restricted in a narrow brain region or distributed broadly in the brain.
8. Related to the above points, the neuroscientific bases of binding should be addressed further. Whether synchrony of neuronal activities in real time or in a local rhythm of neuronal activities reflected in the field potential play a critical role or not, should be determined. The roles of various neuronal activity rhythms and also how they are generated need to be studied further.
9. Neuronal mechanisms of attention should be clarified further. In particular, I think how top-down attention signals are constructed and how they contribute to choose neuronal assemblies coding particular objects or affairs in neuronal circuits of the brain are enigmatic. I also think that the respective functional roles of each neuromodulator molecule such as norepinephrine and dopamine in the attention systems need to be elucidated more precisely.
10. The neuronal mechanisms of metacognitive functions including the working memory are unclear. Precise and detailed clarifications of neuronal activities underlying these higher functions of the brain are desired, although I think it is difficult to address this aspect in humans. fMRI does not have enough spatiotemporal resolutions, and some technical advancement seems to be necessary to significantly progress human studies.
11. Loop structures in neuronal circuits of the brain need to be better described considering the strength and functions of each loop in different conditions.

12. The inputs and outputs of mirror neurons need to be defined, and how the activities of the mirror neurons are generated and how the activities influence metacognitive functions should be clarified further.
13. Excitatory and inhibitory balances or balances of neuronal activities representing opposite, incompatible or contrasting contents, should be further clarified in various conditions including awakened states, REM sleep and non-REM sleep, and also in psychiatric disorders such as schizophrenia, dissociative disorders, autism spectrum disorders and attention deficit hyperactivity disorders.
14. The definitions, distinctions and neuronal bases of various emotions need further clarification. Interaction among emotions should also be studied further. Some neuronal regions are involved in multiple emotions, although there are likely to be subregions in most nuclei. Further studies elucidating more precise and detailed neuronal bases of each emotion is required. Revised classifications of emotions based on neuronal control mechanisms, might contribute to understanding emotions better, although the subjective nature of many emotions seems to hinder research using animals. As explained in chapter eight, control neuronal systems of emotions are complicated and have not been elucidated sufficiently. However, understanding emotions better is essential for understanding the self and mind; as emotions are likely to be a key factor which distinguish brains from computer systems or AI.
15. The relations and interactions between bodily states and mental states when experiencing emotions should be studied further, which would contribute towards improved health care problems, etc. together with further elucidation of the respective roles of neuromodulator molecules working in both the brain and peripheries of the body.
16. Neuronal mechanisms that choose challenging or gambling behaviors are worth being explored further.
17. More precise clarification of neuronal mechanisms that underly subjective awareness and metacognitive functions such as working memory, inference, etc., which include identification of related neurons, characterization of their activities and interactions among them, are desirable.
18. Functions of the cerebral cortices, the frontal cortices in particular, should be more clearly defined, although their rich inter-regional connections and presumably abstract coding contents would make clarification of respective roles of each cortical area difficult.
19. Strong activities of neuronal assemblies seem to be related to awareness. If so, conscious states presumably accompanied with intense activities of a large number of neurons, such as, during epileptic seizure or during consumption of a stimulant such as cocaine might be worth studying further. Uncontrolled and very strong neuronal activities might cause loss of consciousness. Alternatively, they might be accompanied with some conscious states, but be difficult to remember.

Final Message to Readers

I end here, hoping the readers have enjoyed reading this book and that it could provide some useful information which would contribute towards a better understanding of yourself, mind, consciousness and emotions. I think that such an understanding could improve your decision making and selection of actions as an individual and also in societies. I wonder if this book might have aroused new questions in your mind. I welcome such situations. Having questions and asking for clarification of problems is, I think, a very important starting point to better understand complicated affairs and situations.

Index

2-arachidonoylglycerol (2-AG) 132, 136

A

α frequency 142
aberrant salience hypothesis 153, 161, 165, 166
acetylcholine 26–28, 65, 130, 136, 142–144, 149, 192
action potential 20–27, 32, 35, 50, 60, 62, 63, 65, 69, 71, 79, 92, 118, 119, 128
addiction 69, 131, 133, 136, 137, 152
adenosine tri-phosphate (ATP) 24, 29, 92
adenylyl cyclase 28–31, 50, 127, 128
agranular region 112
all-or none law 23
altruism 13, 131
Alzheimer's disease 152, 162, 163, 164
ambiguous figure 68, 96, 182
amine 27, 28, 30, 93, 96, 126, 129, 130, 135, 136, 142, 143, 159, 184, 192, 193
AMPA receptor 27, 32
amygdala 39–41, 49, 60, 67, 82–84, 86, 102–115, 117–122, 127, 129, 131, 157, 159, 160, 172, 187
amyloid β 163
anandamide 132, 136
anger 101, 102, 104, 105, 107, 112, 113, 115, 117, 118, 121, 134, 172, 175
antagonistic inhibition 42, 52, 93, 161
anterograde amnesia 81
anxiety 101–104, 106, 111–113, 115, 117–122, 128, 129, 133, 136, 140, 147, 148, 151, 158, 165, 172, 175, 178, 181
anxiety disorder 115, 129, 136, 140, 151, 158, 165
Aplysia 30, 31, 35, 74
apolipoprotein-E (ApoE) 163
arousal state 9, 15, 65, 115, 128
arousal system 143, 144, 149
artificial intelligence (AI) 6, 78, 89, 90, 92, 95–99, 175, 176, 184–186, 191, 193, 197
Asperger syndrome 155
ataxia 54, 164
attention 3, 7, 8, 28, 64–69, 71, 91, 94, 98, 102, 104, 107, 109, 113, 117, 121, 122, 127, 128, 130, 136, 147–149, 151, 152, 155–157, 159, 161, 162, 165, 166, 172, 178, 181, 182, 185–187, 189, 192, 194, 196, 197
attention deficit hyperactivity disorder 128, 151, 155–157, 161, 165, 166, 187, 197
auditory system 48
autism spectrum disorder 151, 155–157, 161, 165, 166, 187, 197
awareness 2, 7–9, 15–17, 52, 112, 117, 154, 158, 160, 161, 165, 166, 182, 183, 185, 187–189, 192, 197
axon 19–21, 24, 27, 32, 33, 35, 39, 41–46, 49–51, 53, 55, 56, 65, 76, 93, 107, 109, 120, 127, 129, 131, 132, 135

B

β frequency 65, 141
basal forebrain 130, 136, 143, 149
basal ganglia 39–41, 52, 53, 56, 60, 65, 82, 106–108, 110, 111, 114–120, 127, 131, 134, 153, 163, 164, 187
Bayesian estimation 79, 80, 177
bed nucleus of the stria terminalis (BNST) 102–105, 111, 114, 115, 119, 121, 122, 131
belief 6, 17, 146, 152, 165, 176–178, 180, 186
binding problem 61, 62, 71
binocular rivalry 60, 67, 68, 71
bipolar cell 43, 44
bipolar disorder 151, 158, 159, 161, 165, 166
blindsight 3
body states 54, 101, 121
bottom-up signal 69
brain 3, 4, 6, 7, 11, 16, 17, 19, 23–27, 32, 35, 36, 39, 41, 42, 46–48, 50, 52–54, 56, 59–61, 63–65, 67, 69–71, 74, 75, 79–82, 84–86, 89–99, 102–104, 107, 109–112, 115, 117,

119, 121, 122, 128–137, 139–146, 148, 149, 151, 152, 155–158, 160–162, 164, 165, 174, 175, 179, 182–197
brainstem 3, 40, 49, 51, 54, 55, 89, 103, 111, 118, 119, 127, 128
Brodmann 48
bursting state 143

C

calcium 23–25, 27–35, 50, 83, 84, 112, 132, 157–159
cannabinoids 132, 133, 136
cash memory 89–91, 98
caudate nucleus 115
cell body 19, 20, 41, 49
central nervous system (CNS) 39, 40, 43, 52, 64, 67, 71, 139
central pattern generator 8, 52, 65, 80
central processing unit 89–91, 98
central sulcus 41, 49
cerebellum 33, 39, 53, 56, 60, 65, 76, 81, 129, 191
cerebral cortex 32, 39–42, 45, 48, 50, 59–61, 64–66, 82, 105, 115, 143, 188
chaos 174, 175, 177
chemically-gated channel 23–25, 129
cingulate cortex 108–111, 113, 117, 119, 127, 132, 158, 159
circadian rhythm 94, 119, 141, 144, 148
climbing fiber 33, 34, 76, 77, 85, 90
cochlea 48, 56, 59, 65
cognitive cell theory 61, 70, 71
command neuron 52
competition 6, 16, 17, 67, 71, 75, 79, 92, 97, 99, 105, 173
conditioned reflex 15, 81, 85
conditioning of eye blinking 77
cone 43, 44
conflicts 4, 16, 17, 67, 70, 71, 98, 169–171, 174–176, 178, 180, 183–185, 193
consciousness 1, 2, 4, 6–9, 15–17, 39, 92, 93, 112, 118, 139, 146, 148, 151, 152, 154, 161, 164, 171, 180, 182, 184–195, 197, 198
contrast enhancement 94, 95, 99, 161, 193
convergence 29, 44, 55, 120, 192, 194
critical period 75, 76, 85
cyclic adenosine monophosphate (cAMP) 28–31, 35, 50, 127, 128
cyclic guanosine monophosphate (cGMP) 43

D

δ frequency 142
D1-type dopamine receptor 120
D2-type dopamine receptor 120
decision making 3, 4, 9, 13, 16, 17, 59–61, 70, 71, 74, 79, 85, 91, 93, 94, 96, 98, 99, 104, 106, 107, 112–114, 117, 122, 169–172, 174, 176, 177, 180, 183, 185–187, 189, 193, 198
declarative memory 81, 82, 85, 120
deep brain stimulation 134
delusion 152, 153, 165, 186
dementia 162, 165
demerit 1, 4, 5, 9, 12, 16, 70, 71, 96, 98, 139, 148, 169–171, 173, 175, 178–183, 185, 189, 193
dendrite 19, 20, 50
depolarization 20, 23, 25–27, 30–33, 43, 48, 50, 56, 127, 132
depolarization-stimulated suppression of excitation 132
depolarization-stimulated suppression of inhibition 132
depressant 133, 136
depressive disorder 106, 157, 158, 161, 165
diacylglycerol (DAG) 28, 128
diencephalon 39, 40, 103, 117
direct pathway 116, 117, 120, 127, 135, 163
disgust 101, 104–106, 112, 113, 115, 117, 121, 172, 173, 184
disinhibition 53, 93, 133, 137
dissociative disorder 151, 154, 165, 186, 187, 197
dissociative identity disorder 154, 165, 186, 188, 189, 193
divergence 29, 44, 51, 55, 120, 192, 194
dopamine 26, 28, 53, 79, 85, 96, 108, 109, 115, 116, 119–121, 126–129, 133, 135–137, 143, 144, 149, 153, 154, 156, 157, 161, 163–166, 193, 196
dorsal pathway 46, 47
dorsal root ganglia 39, 49
dream 17, 23, 139, 144–149, 154, 183, 184, 186, 188, 189, 191, 193, 195
dualism 17
dynorphin 131, 136
dysgranular region 112

E

electroencephalogram 118, 140, 141, 145, 148, 149
emergence of consciousness 15, 17, 184, 185
emotions 1, 3, 12, 16, 17, 28, 39, 60, 64, 66, 67, 69–71, 79, 92, 93, 97, 99, 101–104, 107, 108, 110–115, 117, 118, 120–122, 126, 129, 134, 135, 140, 152, 155, 158–161, 165, 166, 172–178, 180–189, 191–195, 197, 198
endorphin 131, 136
endplate potential (EPP) 27

engram 82–86, 104
enkephalin 50, 69, 126, 131, 136
epilepsy 81, 152, 160, 161, 188
episodic memory 81, 85, 112, 140
equilibrium potential 21, 25, 26, 35
error signal 33, 34, 76, 77, 79, 85
evolution 11, 14–17, 94, 105, 106, 171, 182, 191, 193, 194
excitation and inhibition balance 161, 166, 187, 193, 195
excitatory postsynaptic potential (EPSP) 25–27, 30–33, 35, 92, 93
excitatory synapse 26, 34, 41, 49, 53
external segment of the globus pallidus 116, 120

F

facial expression 101, 106, 108, 121, 184
fact 1, 8, 9, 81, 85, 92, 104, 174, 176–178, 180, 188, 191, 193–195
fear 82–86, 101–106, 110–115, 117–122, 131, 136, 147, 148, 159, 160, 172, 175
fear conditioning 82–86, 102, 104, 114, 159
feedback 48, 65, 66, 76–78, 80, 85, 108, 111, 174
filial imprinting 9
filling-in 94, 95, 99, 193, 195
fMRI 7, 59, 60, 66, 95, 107, 109, 110, 113, 145, 149, 153, 159, 160, 196
free nerve ending 49

G

γ wave 65, 71
GABA 26–28, 35, 119, 127, 130–133, 137, 158, 161, 166
GABA receptor 27, 28, 35, 132, 158
game theory 171, 172, 178, 181
ganglion cell 43–46, 55
gate 22, 23
genetic risk factor 152, 155, 157, 163
glial cells 20, 126, 132, 194
globus pallidus 52, 53, 115, 116, 120, 127, 128
glutamate 25–28, 32–35, 43, 76, 119, 127, 131–133, 153, 161, 166
glutamate receptor 25, 27, 28, 32–35, 43, 132, 133, 153
G-protein 27, 28, 56
G-protein coupled receptors (GPCRs) 28, 56
granule cell 112

H

habenula 119
habituation 30, 74
hair cell 48, 56, 59, 65
hallucination 152, 153, 161, 162, 165, 166, 186, 187
hard problem 6, 9, 17
Hebb's rule 30, 32
hedonic reaction 108, 131, 136
higher order motor cortex 69
higher order visual cortex 56
hippocampus 32, 39–41, 60, 67, 81–85, 102–104, 109, 111, 112, 114, 117, 118, 120, 121, 127, 129, 131, 140, 153, 157, 159, 162, 187
histamine 26, 28, 126, 130, 135, 136, 143, 149
horizontal cell 44, 94
Huntington's disease 53, 162, 164
hyperpolarization 26, 28, 43, 127
hypothalamic-pituitary-adrenocortical (HPA) axis 55
hypothalamus 39, 41, 49, 54–56, 102–105, 107–111, 113, 115, 117–122, 129–131, 136, 141, 144, 148, 149, 157

I

illusion 4, 95, 96, 99, 176
immediate early gene 82, 84, 86
indirect pathway 115–117, 120, 127, 135, 163, 164
inferior colliculus 39, 48, 56
inhibitory postsynaptic potential (IPSP) 26, 35, 65, 93
inhibitory synapse 26, 34, 36, 45, 52, 77
inositoltriphosphate (IP3) 28, 29, 128
instinctive behavior 15, 16, 80
instructive signal 76, 85
insular cortex 50, 51, 54, 70, 103, 106–109, 111–113, 119, 121, 122, 158, 159
internal segment of the globus pallidus 115–117, 120
intracellular signaling cascade 29, 30, 43, 55, 93, 99
intrinsically active neuron 23, 85, 196
ion channel 20–33, 35, 36, 43, 48–51, 54, 83, 111, 129, 134, 135, 157–159, 161

L

lateral geniculate body 43, 45, 47, 48, 55
lateral inhibition 44, 45, 50, 52, 55, 68, 93–95, 99, 161, 187, 192–194
levels of sleep 141
lewy body 162, 163
life 1, 8, 11–13, 17, 36, 84, 98, 118, 152, 159, 160, 169–171, 185, 193
liking 108, 121, 131, 136
limbic system 111, 112
line connection 96

locus coeruleus 55, 103, 117, 128, 129, 132, 135, 136, 143
long-term depression (LTD) 32–34, 36, 54, 75, 77, 82, 85, 86, 90, 129
long-term effect 4, 16, 70, 98, 169, 170, 176, 178, 179, 181–183, 185
long-term memory 63, 64, 81, 82, 85, 90
long-term potentiation (LTP) 32, 33, 36, 75, 82, 129
loop 29, 45, 48, 53, 56, 65, 66, 93, 99, 111, 115, 120, 122, 185, 192–194, 196

M

manic 158, 159
Mauthner cell 52
means 1, 2, 9, 14, 15, 82, 102, 156, 173, 175, 192
mechanosensitive ion channel 48, 49, 54, 135
medulla oblongata 8, 39, 40, 42, 49, 52, 53, 56, 65, 102, 119, 129
membrane potential 20–26, 32, 35, 43, 118
memory 9, 16, 17, 29, 30, 32, 36, 39, 42, 61, 63–67, 69–72, 74, 80–86, 89–91, 93, 98, 104, 112–114, 117, 120, 129, 140, 145, 148, 149, 154, 155, 161, 163, 165, 172, 174, 180, 183, 184, 186–188, 192–197
merit 1, 2, 4, 5, 9, 11–14, 16, 70, 71, 80, 93, 98, 139, 156, 169–173, 175, 176, 178–183, 185, 189, 192, 193
mesencephalon 39, 40
messenger ribonucleic acid (mRNA) 29, 35, 131
metacognition 4, 9, 15–17, 70, 71, 90, 160, 162, 171, 187
mind 1, 2, 5, 6, 8, 9, 13, 15–17, 39, 66, 69, 92, 93, 109, 146, 151, 152, 160, 165, 171, 176, 178, 182–195, 197, 198
mirror neuron 69, 70, 197
motor control 3, 39, 42, 52, 53, 55, 56, 115, 127, 135
motor learning 33, 54, 76, 77, 81–85, 129
multiple selves 186, 188–190, 193, 195
muscle spindle 41, 42, 49, 52
myelencephalon 39, 40
myelin 20, 21

N

neural correlates of consciousness 7, 9
neural tube 39, 40, 43, 80
neurodegenerative disorder 151
neurodevelopmental disorder 80, 151, 152, 155, 156, 161, 162, 165, 166, 187, 196
neuromodulator 26–28, 50, 66, 69, 92, 93, 96, 99, 109, 126, 130, 134–137, 143, 144, 148, 157, 158, 161, 162, 165, 166, 172, 182–187, 192–197
neuron 8, 9, 19–21, 23–27, 30, 31, 33–36, 41, 42, 44–46, 49, 50, 52, 53, 56, 59–62, 64, 65, 69, 70, 76, 77, 79, 82, 85, 90, 93, 116, 118, 120, 132, 136, 164, 195, 196
neuronal assemblies 61–68, 70–72, 84, 86, 147, 149, 156, 161, 182–184, 186–189, 192, 194, 196, 197
neuronal circuits 8, 34, 35, 39, 48, 52, 53, 64, 66, 70–72, 74–76, 80, 85, 91, 93–96, 99, 103–106, 110, 111, 126, 134, 135, 160, 161, 174, 184, 187, 191–196
neuronal ensemble 83, 84
neurotransmitter 20, 24–30, 35, 36, 48, 50, 76, 79, 93, 118, 119, 127, 130, 132, 136, 162, 195, 196
nitric oxide 126, 133
NMDA receptor 27, 32, 33, 153
nociception 107, 129, 136
non-declarative memory 81, 85
non-REM sleep 141–146, 148, 149, 186, 197
norepinephrine 26, 28, 55, 66, 96, 119, 126, 128, 129, 133, 135, 136, 143, 144, 147–149, 157, 158, 162, 193, 196
nucleus accumbens 104, 106, 108–111, 113–115, 117, 119, 122, 126–128, 131, 133, 135, 136, 153, 158, 172

O

olfactory bulb 39, 50, 56
olfactory epithelium 50, 56
olfactory system 50
opinion 170, 176–181, 191, 193–195
optic chiasma 45, 46, 141, 188
optogenetic 51, 54, 82, 83, 86, 103, 106, 128, 131, 134, 136, 137, 144
orbitofrontal cortex 108, 112–114, 122
orexin 144, 149, 192
oxytocin 109, 118, 126, 131, 132, 136, 160

P

pallidum 106, 108, 115–117, 122, 134
parallel fiber 33, 34, 76, 77
parallel information processing 63
Paramecium 8, 9, 16
parasympathetic nerve 54
Parkinson's disease 53, 108, 128, 129, 134, 135, 152, 162–164
perception 3, 4, 7, 8, 14, 23, 24, 39, 42, 46–48, 50, 59, 61, 65–69, 71, 84, 86, 94, 95, 106, 111–113, 145, 149, 152, 154, 158, 165, 183, 186, 187, 192–194

periaqueductal gray (PAG) 102, 104, 119
peripheral nervous system (PNS) 39
phantom pain 107
phospholipase C 28, 29, 34, 128
photoreceptor 43, 44, 55, 59, 63
physical neuromodulation 134, 135, 158
pleasure 79, 80, 101, 102, 104, 107–113, 117, 120, 121, 127, 131, 133–136, 152, 173, 184
pons 39, 40, 118, 120, 129
population coding 61, 70, 71
pore 22, 23, 25, 27, 33, 35, 51
post-traumatic stress disorder (PTSD) 151, 159
potassium ion channel 23, 29–31
prefrontal cortex 64, 66, 67, 69, 102–105, 113–115, 117, 119–122, 127, 128, 132, 136, 157, 159
primary motor cortex 41, 48, 49, 53, 56, 60
primary visual cortex 3, 40, 41, 45–48, 55, 65, 74–76
prion disease 162, 164
protein kinase 28, 29, 34
proton ion channel 24, 51
psychiatric disorder 17, 115, 136, 151, 152, 157, 161, 165, 166, 186, 191, 193, 195, 197
psychopathy 97, 115, 151, 160
Purkinje cell 34
purpose 11, 17, 77, 92, 93, 98, 99, 134, 137, 139, 175, 185, 193, 195
purposes of sleep 139
putamen 107, 115, 116
pyramidal neuron 45, 47, 53, 120

Q

qualia 3, 4, 6, 8, 9, 15, 17, 43, 59, 63, 64, 66, 70, 71, 108, 153, 158, 159

R

RAM 89–91
raphe nuclei 103, 115, 117, 119, 129, 143
reasoning 176–178, 180
receptive field 45–47, 56
receptor 16, 24, 25–36, 43, 49–51, 54–56, 59, 69, 109, 111, 120, 126–136, 153, 157, 158, 161, 165, 183, 196
recognition 4–9, 13, 14, 23, 39, 48, 59–61, 66–71, 92, 94–96, 99, 114, 140, 165, 182, 183, 186, 187, 189, 192–194
reflex 3, 8, 14–17, 30, 31, 41, 42, 52, 54–56, 74, 77, 81, 85, 90, 171, 180
refractory period 23, 65, 118
reinforcement learning 74, 75, 78–80, 85, 90, 108, 121, 127, 135
releaser 15
REM sleep 119, 141–146, 148, 149, 154, 183, 184, 186, 188, 189, 193, 195, 197
reproduction 13, 14, 118, 139, 173
retina 3, 40, 42–46, 52, 55, 63, 77, 80, 93, 94, 141
reward prediction error 79, 127, 135
reward signal 78–80, 85
rhodopsin 43
rhombencephalon 39, 40
rhythm 23, 52, 64–66, 71, 80, 94, 113, 119, 130, 141, 144, 148, 196
rod 43, 44, 83
ROM 89–91

S

savant syndrome 155
schadenfreude 110
schizophrenia 80, 128, 129, 151–154, 157, 161, 165, 166, 186–188, 193, 197
second messenger 28, 29
seizure 52, 160, 161, 197
selective permeability 21, 22, 35
self 1, 2, 4–6, 9, 12, 13, 17, 39, 74, 75, 85, 93, 102, 107, 108, 110, 112, 121, 130, 136, 146, 148, 149, 152, 154, 159, 160, 165, 166, 171, 172, 176, 182–189, 191–193, 197
self-interest 13
self-organization 74, 75, 85, 130, 136
self-recognition 4–6, 9, 165
self-stimulation 107, 108
semantic memory 81, 85
sensation 24, 28, 42, 49–51, 59, 60, 67–69, 71, 90, 107, 111, 112, 119, 152, 165, 183, 186, 188, 192
sensitization 30, 31, 35, 74
septum 102, 103, 105, 115, 121, 130
sequential information processing 64, 66
serotonin 26, 28, 30, 31, 126, 129, 135, 136, 143, 144, 149, 157, 158, 160
serotonin reuptake inhibitor 129, 136, 158
sham rage 105
short-term effect 98, 170, 178, 181
short-term memory 63, 64, 71, 81, 90
sleep-promoting system 141, 144
sodium ion channel 20, 22, 23, 51
sodium-calcium exchanger 24
sodium-potassium pump 24
somatosensory system 49
sparse coding 61, 71
spike timing dependent plasticity (STDP) 32, 36
spinal cord 3, 8, 19, 25, 39–42, 49, 52, 53, 56, 69, 80, 107, 119, 129–131
split brain 188–190
spreading factor 96, 99
steroids 55, 126, 132, 133, 159

stimulant 133, 136, 197
stress 55, 103, 106, 115, 118, 119, 127–131, 136, 151, 157–159
stretch reflex 3, 8, 15, 41, 42
striatum 53, 104, 106, 108, 110, 112, 114–117, 120, 127, 128, 135, 163, 164
subjective existences 1, 9
substantia nigra 39, 53, 103, 115–117, 119, 120, 126–128, 132, 135, 162, 163
substantia nigra pars reticulate 115–117
subthalamic nucleus 115, 116, 120, 127, 134
superior colliculus 39, 45–47
supervised learning 74–76, 78, 80, 85, 89
suprachiasmatic nuclei 141, 144, 148
survival 9, 11–14, 17, 93, 97–99, 102, 105, 107, 118, 121, 139, 172, 173, 176, 179, 180, 182, 185, 189, 192, 193
Sylvian fissure 41, 50, 112
sympathetic nerve 54
synapse 19, 20, 24–27, 30–36, 41–45, 48, 49, 52, 53, 75–77, 83, 85, 93, 94, 96, 120, 130, 131, 157, 174, 192, 194
synaptic plasticity 27, 29, 30, 32, 34–36, 75, 77, 79, 82, 93, 108, 127, 129, 133, 135, 158, 191–195
synaptic vesicle 24, 25, 65

T

θ frequency 65, 141
taste system 51
tegmental nuclei 119, 130, 136, 143, 149
telencephalon 39, 40, 80, 114
thalamus 39, 41, 43–51, 53, 55, 56, 65, 67, 80, 103, 104, 111–120, 127, 134, 143, 163, 187
theory of mind 5, 6, 13, 15, 69, 109, 160, 184
threshold 20, 23, 29, 35, 107
tonically active state 143
tonotopy 49
top-down signal 50, 66, 67, 69, 71, 189
transcranial direct current stimulation 134, 158
transcranial magnetic stimulation 134, 135, 158
transcranial ultrasound stimulation 134
transcription 29, 35, 55, 83, 118, 132, 141, 161
transcutaneous electrical nerve stimulation 134
transducin 43
transient receptor potential (TRP) channel 24
trial and error learning 80
tuberomamillary nucleus 130, 136

U

unihemispheric sleep 145, 146, 189
unity of consciousness 188

V

ventral pathway 46, 47, 63
ventral tegmental area 79, 103, 104, 109, 115, 117, 119, 126, 127, 131, 135, 136, 153, 165
vestibulo-ocular reflex (VOR) 3, 8, 15, 42, 54, 77, 81
visual agnosia 4
visual system 43, 45, 47
voltage-gated channel 22, 23, 25, 28, 30, 31, 157
volume transmission 28, 93, 126, 129–131, 192
voluntary behavior 15, 17, 63

W

wanting 108, 121, 127, 133, 135, 136
working memory 63–67, 70–72, 81, 84, 90, 91, 93, 98, 113, 114, 117, 180, 192, 194, 196, 197